UNIVERSITY MATHEMATICAL TEXTS
EDITORS/ALAN JEFFREY AND IAIN T. ADAMSON

MATHEMATICAL ANALYSIS

RECENT UNIVERSITY MATHEMATICAL TEXTS

An Introduction to Modern Mathematics/Albert Monjallon
Quantum Mechanics/R. A. Newing and J. Cunningham
Probability/J. R. Gray
Series Expansions for Mathematical Physicists/H. Meschkowski
Elementary Rings and Modules/Iain T. Adamson
An Introduction to the Theory of Statistics/R. L. Plackett

GORDON H. FULLERTON
LECTURER IN MATHEMATICS
UNIVERSITY OF NOTTINGHAM

MATHEMATICAL ANALYSIS

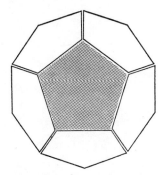

OLIVER & BOYD EDINBURGH

OLIVER AND BOYD
Tweeddale Court
14 High Street
Edinburgh EH1 1YL
A Division of Longman Group Limited
UMT 41
0 05 002346 2
First published 1971
© 1971 Gordon H. Fullerton
All rights reserved
No part of this book may be reproduced, stored in a
retrieval system, or transmitted in any form or by any means—
electronic, mechanical, photocopying, recording or otherwise—
without the prior permission of the Copyright Owner and the
Publisher. All requests should be addressed to the Publisher
in the first instance.

Printed in Great Britain by
R. & R. Clark Ltd., Edinburgh

PREFACE

This book is based on lectures which I have given to second-year undergraduates at the University of Nottingham. It assumes a familiarity with the material currently discussed in first-year analysis courses at British universities, but is otherwise self-contained. I have tried throughout to present the basic ideas of abstract analysis in an informal style, while maintaining the usual standards of mathematical rigour. For example, I have illustrated the proofs of several of the harder theorems graphically, to give the reader an insight into why the results are valid. I have emphasized the essential nature of the hypotheses of most of the stated theorems by appropriate counterexamples in the text or in the problems at the end of each chapter. These problems form an important part of the book, and I strongly recommend the reader to work through as many of them as he can.

In Chapter 1, the concepts of compactness, connectedness and completeness are introduced in a metric space setting. Chapter 2 is concerned with properties of continuous functions between metric spaces. Uniform convergence of sequences of functions is defined and related to the completeness of $C(X)$. Proofs of the Stone-Weierstrass and Ascoli theorems are given. The possibility of extending much of the above material to a general topological setting is briefly indicated. Chapter 3 deals with further results on the uniform convergence of sequences and series of functions. In Chapter 4, the Lebesgue integral is defined, using Daniell's approach, as a linear functional on a certain class of functions. The relation to the traditional measure-theoretic definition is clearly explained. A discussion of the L^p spaces and double integrals is included. Chapter 5 contains the basic results of the L^1 and L^2 theories of Fourier transforms.

I wish to express my gratitude to Dr. Iain T. Adamson for his helpful comments on a preliminary draft of this book. I wish also to thank my wife, Angela, for typing the manuscript and for her patience and encouragement during the entire project.

Gordon H. Fullerton

Nottingham,
November 1970.

CONTENTS

CHAPTER 1   **METRIC SPACES**
                    1   Metric and normed spaces   **1**
                    2   Open and closed sets   **4**
                    3   Compactness   **10**
                    4   Connectedness   **17**
                    5   Convergence   **22**
                    6   Consequences of completeness   **26**
                         Problems 1   **28**

CHAPTER 2   **CONTINUOUS FUNCTIONS**
                    7   Definition and topological conditions   **32**
                    8   Preservation of compactness and connectedness   **34**
                    9   Uniform convergence   **39**
                    10   Uniform continuity   **44**
                    11   Weierstrass's Theorem   **46**
                    12   The Stone–Weierstrass Theorem   **50**
                    13   Compactness in $C(X)$ **55**
                    14   Topological spaces: an aside   **58**
                         Problems 2   **59**

CHAPTER 3   **FURTHER RESULTS ON UNIFORM CONVERGENCE**
                    15   Uniform convergence and integration   **63**
                    16   Uniform convergence and differentiation   **68**
                    17   Uniform convergence of series   **70**
                    18   Tests for uniform convergence of series   **71**
                    19   Power series   **75**
                         Problems 3   **80**

CHAPTER 4   **LEBESGUE INTEGRATION**
                    20   The collection K and null sets   **85**
                    21   The Lebesgue integral   **91**
                    22   Convergence theorems   **97**
                    23   Relation between Riemann and
                         Lebesgue integration   **102**
                    24   Daniell integrals   **109**
                    25   Measurable functions and sets   **113**
                    26   Complex-valued functions: L^p spaces   **120**
                    27   Double integrals   **127**
                         Problems 4   **133**

CHAPTER 5   **FOURIER TRANSFORMS**
                    28   L^1 theory: elementary results   **138**
                    29   The inversion theorem   **140**
                    30   L^2 theory   **144**
                         Problems 5   **148**

                         Index   **150**

CHAPTER 1
METRIC SPACES

§1. Metric and normed spaces

Analysis could be described very crudely as the branch of mathematics dealing with approximations. The natural setting in which to discuss approximations is a metric space, i.e. a space in which there is a concept of distance.

1.1 DEFINITION. A **metric space** is a pair (X, d) consisting of a non-empty set X and a real-valued function d on $X \times X$ satisfying the following conditions:

(a) $d(x, y) \geq 0$ for all x, y in X with equality if and only if $x = y$;
(b) $d(x, y) = d(y, x)$ for all x, y in X;
(c) $d(x, y) \leq d(x, z) + d(z, y)$ for all x, y, z in X.

A function d on X satisfying (a), (b) and (c) is called a **metric** on X.

We naturally think of $d(x, y)$ as the "distance" from x to y. Conditions (a), (b) and (c) are simply abstractions obtained from properties of distance in space. For example, (c) corresponds to the property that the length of one side of a triangle is less than or equal to the sum of the lengths of the other two sides. For this reason, (c) is often called the "triangle inequality".

1.2 *Examples.* (a) $X = \mathbf{C}$, the set of complex numbers; d, the **usual metric** on $\mathbf{C}$, defined by

$$d(x, y) = |x - y| \qquad (x, y \in \mathbf{C}).$$

(b) $X = \mathbf{R}^k$, the set of k-tuples of real numbers, k being a positive integer; d, the **usual metric** on $\mathbf{R}^k$, defined by

$$d(x, y) = \left\{ \sum_{i=1}^{k} (x_i - y_i)^2 \right\}^{\frac{1}{2}} \quad (x = (x_1, \ldots, x_k), y = (y_1, \ldots, y_k) \in \mathbf{R}^k).$$

To show that d satisfies the triangle inequality, we first note that if $a_1, \ldots, a_k, b_1, \ldots, b_k \in \mathbf{R}$, we have

$(a_1^2 + \ldots + a_k^2)(b_1^2 + \ldots + b_k^2)$

$$= a_1^2 b_1^2 + \ldots + a_k^2 b_k^2 + \sum_{\substack{i,j=1 \\ i \neq j}}^{k} a_i^2 b_j^2$$

$$= \left\{ \sum_{i=1}^{k} a_i b_i \right\}^2 + \sum_{\substack{i,j=1 \\ i \neq j}}^{k} (a_i^2 b_j^2 - a_i b_i a_j b_j)$$

$$= \left\{ \sum_{i=1}^{k} a_i b_i \right\}^2 + \sum_{i<j} (a_i b_j - a_j b_i)^2,$$

so that

$$\left| \sum_{i=1}^{k} a_i b_i \right| \leq \left\{ \sum_{i=1}^{k} a_i^2 \right\}^{\frac{1}{2}} \left\{ \sum_{i=1}^{k} b_i^2 \right\}^{\frac{1}{2}} \qquad \text{(Cauchy's inequality)}.$$

Hence, if $x = (x_1, \ldots, x_k)$, $y = (y_1, \ldots, y_k)$ and $z = (z_1, \ldots, z_k)$ belong to $\mathbf{R}^k$, we have

$\{d(x, z) + d(z, y)\}^2$

$$= \sum_{i=1}^{k} (x_i - z_i)^2 + \sum_{i=1}^{k} (z_i - y_i)^2 + 2 \left\{ \sum_{i=1}^{k} (x_i - z_i)^2 \cdot \sum_{i=1}^{k} (z_i - y_i)^2 \right\}^{\frac{1}{2}}$$

$$\geq \sum_{i=1}^{k} (x_i - z_i)^2 + \sum_{i=1}^{k} (z_i - y_i)^2 + 2 \sum_{i=1}^{k} (x_i - z_i)(z_i - y_i)$$

$$= \sum_{i=1}^{k} \{(x_i - z_i) + (z_i - y_i)\}^2$$

$$= d(x, y)^2,$$

so that $d(x, y) \leq d(x, z) + d(z, y)$.

(c) If d is a metric on a set X and A is a non-empty subset of X, then the function d_A defined by

$$d_A(x, y) = d(x, y) \qquad (x, y \in A)$$

is a metric on A. d_A is called the **relative metric** induced by d on A. The metric space (A, d_A) is called a **subspace** of (X, d). For example,

$\mathbf{R} = \mathbf{R}^1$, with its usual metric, is a subspace of $\mathbf{C}$, with its usual metric.

(d) $X = \mathbf{m}$, the set of all bounded sequences of complex numbers; d defined by
$$d(x, y) = \sup_{n \geq 1} |x_n - y_n| \quad (x = (x_n),\ y = (y_n) \in \mathbf{m}).$$

(e) $X = \mathbf{s}$, the set of all sequences of complex numbers; d defined by
$$d(x, y) = \sum_{n=1}^{\infty} \frac{|x_n - y_n|}{2^n(1 + |x_n - y_n|)} \quad (x = (x_n),\ y = (y_n) \in \mathbf{s}).$$

To show that d satisfies the triangle inequality, we first note that if a, b, c are non-negative real numbers and $c \leq a+b$, then
$$\frac{c}{1+c} = 1 - \frac{1}{1+c} \leq 1 - \frac{1}{1+a+b}$$
$$= \frac{a+b}{1+a+b} \leq \frac{a}{1+a} + \frac{b}{1+b}.$$

Hence, if $x = (x_n)$, $y = (y_n)$ and $z = (z_n)$ belong to $\mathbf{s}$, we have
$$d(x, z) + d(z, y) = \sum_{n=1}^{\infty} \frac{1}{2^n}\left[\frac{|x_n - z_n|}{1 + |x_n - z_n|} + \frac{|z_n - y_n|}{1 + |z_n - y_n|}\right]$$
$$\geq \sum_{n=1}^{\infty} \frac{|x_n - y_n|}{2^n(1 + |x_n - y_n|)}$$
$$= d(x, y).$$

(f) $X = C[a, b]$, the set of continuous complex-valued functions on the closed interval $[a, b]$; d, the **uniform metric** on $[a, b]$, defined by
$$d(x, y) = \sup_{a \leq t \leq b} |x(t) - y(t)| \quad (x, y \in C[a, b]).$$

(g) $X = C[a, b]$; d defined by
$$d(x, y) = \int_a^b |x(t) - y(t)|\, dt \quad (x, y \in C[a, b]).$$

Examples 1.2 (f) and (g) show that there may be more than one interesting metric on a set, and thus emphasize that a metric space is a composite object consisting of a non-empty set X and a metric d on X. Nevertheless, when it is unlikely to cause confusion, we shall sometimes denote a metric space by the same symbol as the underlying set.

Many interesting metric spaces, including most of the examples mentioned above, are in fact normed spaces.

1.3 DEFINITION. A **norm** on a vector space X over the real or complex field is a real-valued function $\|\cdot\|$ on X satisfying the following conditions:

(a) $\|x\| \geqslant 0$ for each x in X with equality if and only if $x = 0$;
(b) $\|\alpha x\| = |\alpha| \|x\|$ for each x in X and each scalar α;
(c) $\|x+y\| \leqslant \|x\| + \|y\|$ for all x, y in X.

Let $\|\cdot\|$ be a norm on a vector space X, and define

(1.1) $$d(x, y) = \|x-y\| \qquad (x, y \in X).$$

By (a), $d(x, y) \geqslant 0$ with equality if and only if $x = y$; by (b), $d(x, y) = d(y, x)$; finally, replacing x by $x-z$ and y by $z-y$ in (c), we obtain

$$d(x, y) = \|x-y\| \leqslant \|x-z\| + \|z-y\| = d(x, z) + d(z, y).$$

Thus d is a metric on X.

1.4 DEFINITION. A **normed space** is a metric space (X, d) in which X is a vector space over the real or complex field and d is given by (1.1) for some norm $\|\cdot\|$ on X.

With the exception of 1.2(c), each of the sets in the above examples is a vector space under the usual algebraic operations. Except in 1.2(e), each of the metrics on these vector spaces comes from a norm. For instance in 1.2(f)

$$d(x, y) = \|x-y\| \quad \text{where} \quad \|x\| = \sup_{a \leqslant t \leqslant b} |x(t)|.$$

(This norm is called the **uniform norm** on $[a, b]$.) The metric d on **s** does not come from a norm since if $x, y \in \mathbf{s}$ and $\alpha \in \mathbf{C}$, $d(\alpha x, \alpha y)$ and $|\alpha| d(x, y)$ are in general different.

§2. Open and closed sets

Many of the deepest theorems of analysis depend on certain topological concepts. In the next few sections, we shall introduce these concepts and derive some fundamental results concerning them.

2.1 DEFINITION. A subset G of a metric space X is said to be **open in** X (or simply **open**) if, for each point y in G, there is a positive real number r such that $\{x \in X : d(x, y) < r\}$ is contained in G.

2.2 *Examples.* (a) If X is any metric space, then X and $\emptyset$, the empty set, are open in X.

(b) $G = \{x \in \mathbf{R} : 0 < x < 1\}$ is open in $\mathbf{R}$;
$H = \{x \in \mathbf{R} : 2 \leqslant x \leqslant 3\}$ is not open in $\mathbf{R}$.

If $y \in G$ and $r = \min\{y, 1-y\}$, then $r > 0$ and $\{x \in \mathbf{R} : |x-y| < r\}$ is contained in G; $2 \in H$, but there is no real number $s > 0$ such that $\{x \in \mathbf{R} : |x-2| < s\}$ is contained in H.

(c) $G_1 = \{z \in \mathbf{C} : 2 < \operatorname{Re} z < 3\}$ is open in $\mathbf{C}$;
$H_1 = \{z \in \mathbf{C} : 0 < \operatorname{Re} z < 1 \text{ and } \operatorname{Im} z = 0\}$ is not open in $\mathbf{C}$.

If $w \in G_1$ and $r = \min\{\operatorname{Re} w - 2, 3 - \operatorname{Re} w\}$, then $r > 0$ and $\{z \in \mathbf{C} : |z-w| < r\} \subseteq G_1$; $\frac{1}{2} \in H_1$, but there is no real number $s > 0$ such that $\{z \in \mathbf{C} : |z-\frac{1}{2}| < s\} \subseteq H_1$.

(d) If X is a metric space, $x_0 \in X$ and $\rho > 0$, then
$$B = \{x \in X : d(x, x_0) < \rho\} \text{ is open in } X.$$
For let $y \in B$, and define $r = \rho - d(y, x_0)$. Then $r > 0$ and, if $x \in X$ and $d(x, y) < r$, we have

$$\begin{aligned} d(x, x_0) &\leqslant d(x, y) + d(y, x_0) \quad \text{by the triangle inequality} \\ &< r + d(y, x_0) \\ &= \rho, \end{aligned}$$

so that $\{x \in X : d(x, y) < r\} \subseteq B$.

The argument is illustrated in Figure 2.1. Although implicitly assuming that $X = \mathbf{R}^2$, and so to a certain extent misleading, this picture does suggest the proof. Even in more complicated situations, pictures of this sort are often helpful.

The set B in Example 2.2(d) is called the **open ball** in X having **centre** x_0 and **radius** ρ and denoted by $B(x_0; \rho)$.

According to Examples 2.2(b) and (c), the set
$$\{z \in \mathbf{C} : 0 < \operatorname{Re} z < 1 \text{ and } \operatorname{Im} z = 0\}$$
is not open in $\mathbf{C}$, but is an open subset of $\mathbf{R}$. Thus openness is not

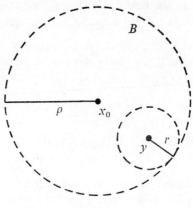

Figure 2.1

an intrinsic property of a set E: it depends on the metric space in which E is embedded.

Our first theorem lists the important properties of open sets.

2.3 THEOREM. *Let X be a metric space. Then* (a) *X and $\emptyset$ are open subsets of X;* (b) *the intersection of a finite collection of open subsets of X is an open subset of X;* (c) *the union of any collection of open subsets of X is an open subset of X.*

Proof. (a) These facts were pointed out above in Example 2.2(a).

(b) Let $\{G_1, G_2, \ldots, G_n\}$ be a finite collection of open subsets of X, and write $H = \bigcap_{i=1}^{n} G_i$. Let $y \in H$. Then, for $i = 1, 2, \ldots, n$, we have $y \in G_i$ and, since G_i is open, there is a positive real number r_i such that $B(y; r_i) \subseteq G_i$. Write $r = \min\{r_1, r_2, \ldots, r_n\}$. Then $r > 0$ and $B(y; r) \subseteq B(y; r_i) \subseteq G_i$ for $i = 1, 2, \ldots, n$, so that $B(y; r) \subseteq H$. Thus H is an open subset of X.

(c) Let $\{G_i : i \in I\}$ be an arbitrary collection of open subsets of X, and write $H = \bigcup_{i \in I} G_i$. Let $y \in H$. Then there is an index i_0 in I such that $y \in G_{i_0}$. Since G_{i_0} is open, there is an open ball B, with centre y, contained in G_{i_0}. Since $G_{i_0} \subseteq H$, B is contained in H and H is open.

We note that the intersection of an infinite collection of open subsets of a metric space need not be open. For example,

$$G_n = \left\{ x \in \mathbf{R} : -\frac{1}{n} < x < 1 + \frac{1}{n} \right\}$$

METRIC SPACES 7

is an open subset of **R** for each positive integer n, but
$$\bigcap_{n=1}^{\infty} G_n = \{x \in \mathbf{R}: 0 \leqslant x \leqslant 1\}$$
is not open in **R**.

Let X be a metric space and Y be a non-empty subset of X. We have already pointed out that a subset E of Y may be open in Y (considered as a subspace of X) without being open in X. Our next result describes the relationship between open subsets of Y and of X.

2.4 THEOREM. *Let X be a metric space and Y be a non-empty subset of X. Then a subset E of Y is open in Y if and only if there is an open subset G of X such that $E = Y \cap G$.*

Proof. Suppose that E is open in Y. Then, for each $u \in E$, there is a positive real number r_u such that $\{y \in Y: d(y, u) < r_u\} \subseteq E$. Clearly
$$E = \bigcup_{u \in E} \{y \in Y: d(y, u) < r_u\}.$$
Define
$$G = \bigcup_{u \in E} \{x \in X: d(x, u) < r_u\}.$$
Then G, being the union of a collection of open balls in X, is an open subset of X and
$$Y \cap G = \bigcup_{u \in E} \{y \in Y: d(y, u) < r_u\} = E.$$
Conversely, suppose that $E = Y \cap G$ where G is an open subset of X. Let $u \in E$. Then $u \in G$, and so there is a positive real number r such that $\{x \in X: d(x, u) < r\} \subseteq G$. Hence
$$\{y \in Y: d(y, u) < r\} = Y \cap \{x \in X: d(x, u) < r\} \subseteq Y \cap G = E,$$
showing that E is open in Y.

2.5 DEFINITION. A subset F of a metric space X is said to be **closed in** X (or simply **closed**) if its complement, $C(F) = X - F$, is open in X.

2.6 *Examples.* (a) If X is any metric space, then X and $\emptyset$ are closed in X.

These statements follow immediately from Example 2.2(a).

(b) $F = \{x \in \mathbf{R}: 0 \leqslant x \leqslant 1\}$ is closed in **R**.

$C(F)$ is the union of $\{x \in \mathbf{R}: x < 0\}$ and $\{x \in \mathbf{R}: x > 1\}$, each of which is open in **R**.

(c) $F_1 = \{(x, y) \in \mathbf{R}^2 : x^2 + y^2 \leqslant 1\}$ is closed in $\mathbf{R}^2$.
$C(F_1) = \{(x, y) \in \mathbf{R}^2 : x^2 + y^2 > 1\}$, an open subset of $\mathbf{R}^2$.

(d) If X is a metric space, $x_0 \in X$ and $\rho > 0$, then the set $\{x \in X : d(x, x_0) \leqslant \rho\}$ is closed in X. (Verification is left to the reader.) This set is called the **closed ball** in X having **centre** x_0 and **radius** ρ.

We note that a metric space may have subsets which are neither open nor closed. For example, $\{x \in \mathbf{R} : 0 \leqslant x < 1\}$ is neither open nor closed in $\mathbf{R}$. Even more curious, every metric space X has subsets which are both open and closed. For example, X and $\emptyset$ are both open and closed in X. We shall see later that $\mathbf{R}^k$ and $\emptyset$ are the only subsets of $\mathbf{R}^k$ which are both open and closed.

We now state and prove the analogues of 2.3 and 2.4 for closed sets.

2.7 THEOREM. *Let X be a metric space. Then (a) X and $\emptyset$ are closed subsets of X; (b) the union of a finite collection of closed subsets of X is a closed subset of X; (c) the intersection of any collection of closed subsets of X is a closed subset of X.*

Proof. These results can be deduced from the corresponding results in 2.3 using the De Morgan laws concerning complements of unions and intersections. For example to prove (b), let $\{F_1, F_2, \ldots, F_n\}$ be a finite collection of closed subsets of X. Then $\{C(F_1), \ldots, C(F_n)\}$ is a finite collection of open subsets of X, and $C(\bigcup_{i=1}^{n} F_i) = \bigcap_{i=1}^{n} C(F_i)$ is open by 2.3(b). Thus $\bigcup_{i=1}^{n} F_i$ is closed.

2.8 THEOREM. *Let X be a metric space and Y be a non-empty subset of X. Then a subset E of Y is closed in Y if and only if there is a closed subset F of X such that $E = Y \cap F$.*

Proof. E is closed in $Y \Leftrightarrow Y - E$ is open in Y
$\Leftrightarrow$ there is an open subset G of X such that
$Y - E = Y \cap G$
$\Leftrightarrow$ there is an open subset G of X such that
$E = Y - (Y - E)$
$= Y - (Y \cap G) = Y \cap (X - G)$.

The result follows.

We have already remarked that a subset of a metric space may be neither open nor closed. However, we can associate, in a natural way, with each subset A of a metric space a corresponding open set, the interior of A, and a corresponding closed set, the closure of A.

2.9 DEFINITION. Let X be a metric space, A be a subset of X and x_0 be a point of X.

x_0 is said to be an **interior point** of A (in X) if there is an open subset G of X such that $x_0 \in G \subseteq A$: in this case, A is called a **neighbourhood** of x_0. The set of interior points of A is called the **interior** of A and denoted by A^0.

x_0 is said to be a **cluster point** of A (in X) if every neighbourhood of x_0 contains a point of A distinct from x_0. The set of cluster points of A is called the **derived set** of A and denoted by A'. The set $A^- = A \cup A'$ is called the **closure** of A.

2.10 REMARK. (a) x_0 is an interior point of A if and only if there is an open ball, with centre x_0, contained in A.

(b) x_0 is a cluster point of A if and only if for every $\varepsilon > 0$, there is a point x in A such that $0 < d(x, x_0) < \varepsilon$.

2.11 *Examples.* (a) In **R**, $[0, 1]^0 = (0, 1)$ and $[0, 1]^- = [0, 1]' = [0, 1]$.

(b) In **R**, $(0, 1)^0 = (0, 1)$ and $(0, 1)^- = (0, 1)' = [0, 1]$.

(c) In **R**, $A = \{1/n : n \text{ is a positive integer}\}$ has no interior points, i.e. $A^0 = \emptyset$, and $A' = \{0\}$.

2.12 THEOREM. *Let X be a metric space and A be a subset of X. Then*

(a) *A^0 is the largest open subset of X contained in A;*
(b) *A^- is the smallest closed subset of X containing A.*

Proof. (a) If $x \in A^0$, there is an open subset G of X such that $x \in G \subseteq A$. Thus A^0 is contained in the union U of the collection of open subsets of X contained in A. On the other hand, if H is an open subset of X contained in A, then $H \subseteq A^0$. Thus $A^0 = U$ and 2.3 implies that A^0 is open. Clearly A^0 is the largest open subset of X contained in A.

(b) We note that

$$x \in C(A^-) = C(A) \cap C(A') \Leftrightarrow \text{there is an } \varepsilon > 0 \text{ such that}$$
$$B(x; \varepsilon) \cap A = \emptyset$$
$$\Leftrightarrow \text{there is an } \varepsilon > 0 \text{ such that}$$
$$B(x; \varepsilon) \subseteq C(A)$$
$$\Leftrightarrow x \in \bigl(C(A)\bigr)^0.$$

Thus $C(A^-) = (C(A))^0$ is open by (a), and A^- is closed. Let F be a closed subset of X containing A. Then $C(F)$ is open and is contained in $C(A)$. Hence, by (a), $C(F) \subseteq (C(A))^0 = C(A^-)$, which implies that $F \supseteq A^-$.

2.13 COROLLARY. *Let X be a metric space and A be a subset of X. Then*
(a) *A is open if and only if every point of A is an interior point of A;*
(b) *A is closed if and only if every cluster point of A belongs to A.*

§3. Compactness

The concept of compactness is perhaps the most important topological idea we shall discuss. (Loosely speaking, a topological concept is one which can be described in terms of open sets.) Many of the results which we shall prove in later chapters depend on compactness arguments.

3.1 DEFINITION. Let X be a metric space and K be a subset of X. A collection $\{G_i : i \in I\}$ of subsets of X is called an **open covering** of K if each of the sets G_i is open and $\bigcup_{i \in I} G_i \supseteq K$. K is said to be **compact** (in X) if every open covering of K contains a finite subcovering of K.

Thus if K is compact and $\{G_i : i \in I\}$ is an open covering of K, there is a finite subset I_0 of I such that $\bigcup_{i \in I_0} G_i \supseteq K$.

Using 2.4, we show that compactness is an intrinsic property of a set K and the metric on K: it does *not* depend on the metric space in which K is embedded.

3.2 THEOREM. *Let X be a metric space and Y be a non-empty subset of X. Then a subset K of Y is compact in Y if and only if it is compact in X.*

Proof. Suppose first that K is compact in X, and let $\{G_i : i \in I\}$ be a collection of open subsets of Y which covers K. Then, for each $i \in I$, there is by 2.4 an open subset H_i of X such that $G_i = Y \cap H_i$. The collection $\{H_i : i \in I\}$ covers K, and so, since K is assumed to be compact in X, there is a finite subset I_0 of I such that $\bigcup_{i \in I_0} H_i \supseteq K$. Then

$$\bigcup_{i \in I_0} G_i = \bigcup_{i \in I_0} (Y \cap H_i) \supseteq Y \cap K = K.$$

It follows that K is compact in Y.

The proof of the converse is similar, and is left to the reader.

The results about open and closed sets proved in §2 occur in pairs. It is therefore interesting to note that compactness, defined in 3.1 in terms of open sets, could have been defined in terms of closed sets.

3.3 DEFINITION. A collection $\{A_i : i \in I\}$ of sets is said to have the **finite intersection property** if $\bigcap_{i \in I_0} A_i \neq \emptyset$ for every finite subset I_0 of I.

3.4 THEOREM. *A metric space X is compact if and only if the intersection of every collection of closed subsets of X having the finite intersection property is non-empty.*

Proof. Suppose that X is compact, and let $\{F_i : i \in I\}$ be a collection of closed subsets of X having the finite intersection property. We must show that $\bigcap_{i \in I} F_i \neq \emptyset$: we argue by contradiction. Suppose that $\bigcap_{i \in I} F_i = \emptyset$. Then $\{C(F_i) : i \in I\}$ is an open covering of X, and, since X is compact, there is a finite subset I_0 of I such that $\bigcup_{i \in I_0} C(F_i) = X$. But then $\bigcap_{i \in I_0} F_i = \emptyset$, contrary to the hypothesis that $\{F_i : i \in I\}$ has the finite intersection property.

Conversely, suppose that the intersection of every collection of closed subsets of X having the finite intersection property is non-empty. Let $\{G_i : i \in I\}$ be an open covering of X. Then $\{C(G_i) : i \in I\}$ is a collection of closed sets and $\bigcap_{i \in I} C(G_i) = \emptyset$. By our hypothesis, this collection cannot have the finite intersection property; so there is a finite subset I_0 of I such that $\bigcap_{i \in I_0} C(G_i) = \emptyset$. Thus $\bigcup_{i \in I_0} G_i = X$, and it follows that X is compact.

3.5 COROLLARY. *Let (F_n) be a sequence of non-empty closed subsets of a compact metric space X such that $F_{n+1} \subseteq F_n$ ($n = 1, 2, \ldots$). Then $\bigcap_{n=1}^{\infty} F_n \neq \emptyset$.*

Proof. The inclusions $F_{n+1} \subseteq F_n$, $n = 1, 2, \ldots$, imply that the collection $\{F_n : n = 1, 2, \ldots\}$ has the finite intersection property.

3.6 NEAREST POINT THEOREM. *Let K be a compact subset of a metric space X and x_0 be a point of X. Then there is a point y_0 in K such that*

$$d(x_0, y_0) = \inf_{y \in K} d(x_0, y).$$

Proof. Write $m = \inf_{y \in K} d(x_0, y)$ and, for each positive integer n, let

$$F_n = \left\{y \in K : d(x_0, y) \leq m + \frac{1}{n}\right\} = K \cap \left\{y \in X : d(x_0, y) \leq m + \frac{1}{n}\right\}.$$

(See Figure 3.1.) F_n is a non-empty closed subset of the compact metric space K and $F_{n+1} \subseteq F_n$ ($n = 1, 2, \ldots$). Hence, by 3.5, $\bigcap_{n=1}^{\infty} F_n \neq \emptyset$. If $y_0 \in \bigcap_{n=1}^{\infty} F_n$, then, for each positive integer n,

$$m \leq d(x_0, y_0) \leq m + \frac{1}{n},$$

which implies that $d(x_0, y_0) = m$.

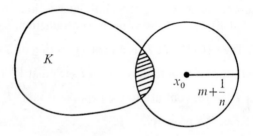

Figure 3.1

We now establish some of the properties of compact sets.

3.7 DEFINITION. A subset E of a metric space X is said to be **bounded** if there is a positive real number M and a point x_0 in X such that $E \subseteq \{x \in X : d(x, x_0) \leq M\}$.

Thus E is bounded if and only if it is contained in some ball in X. If E is contained in the ball $\{x \in X : d(x, x_0) \leq M\}$ having centre x_0 and x_1 is another point of X, then for each $x \in X$, we have

$$d(x, x_1) \leq d(x, x_0) + d(x_0, x_1) \leq M + d(x_0, x_1) = M_1 \text{ say},$$

so that E is contained in the ball $\{x \in X : d(x, x_1) \leq M_1\}$ having centre x_1.

3.8 THEOREM. *Let K be a compact subset of a metric space X. Then K is closed and bounded.*

Proof. To prove that K is closed, we show that $C(K)$ is open. Let $x \in C(K)$. Then, for each $y \in K$, there are open neighbourhoods V_y and W_y of x and y respectively such that $V_y \cap W_y = \emptyset$. For example, we could take V_y, W_y to be open balls having centres x, y and radius less than or equal to $\tfrac{1}{2}d(x, y)$. (See Figure 3.2.) The collection $\{W_y : y \in K\}$ is an open covering of K, and so, since K is compact, there is a finite set of points $y_1, y_2, \ldots, y_n$ in K such that $\bigcup_{i=1}^{n} W_{y_i} \supseteq K$. Then $V = \bigcap_{i=1}^{n} V_{y_i}$ is an open neighbourhood of x and $V \cap K = \emptyset$. Hence $V \subseteq C(K)$ and x is an interior point of $C(K)$. Thus every point of $C(K)$ is an interior point of $C(K)$, and $C(K)$ is open. (The astute reader has perhaps noticed that this result can be deduced easily from 3.6.)

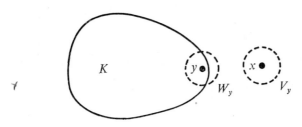

Figure 3.2

To show that K is bounded, let x_0 be any point of X. Then the open balls $B(x_0; n)$, $n = 1, 2, \ldots$, form an open covering of X, and hence of K. Since K is compact and $B(x_0; n+1) \supseteq B(x_0; n)$, $n = 1, 2, \ldots$, there is a positive integer n_0 such that $K \subseteq B(x_0; n_0)$.

3.9 THEOREM. *Let X be a compact metric space and F be a closed subset of X. Then F is compact.*

Proof. Let $\mathscr{G}$ be an open covering of F. Then $\mathscr{H} = \mathscr{G} \cup \{C(F)\}$ is an open covering of X. Since X is compact, $\mathscr{H}$ contains a finite subcovering $\mathscr{H}'$ of X. $\mathscr{H}' - \{C(F)\}$ is then a finite subset of $\mathscr{G}$ and covers F. It follows that F is compact.

According to 3.8, a compact subset of a metric space is closed and bounded. The converse of this result is false in general. For example, consider the subset $E = \{x \in \mathbf{Q} : x > 0 \text{ and } 2 < x^2 < 3\}$ of the metric space $\mathbf{Q}$ of rational numbers, considered as a subspace of $\mathbf{R}$.

E is certainly bounded. Also, since neither $\sqrt{2}$ nor $\sqrt{3}$ is rational, $E = \mathbf{Q} \cap [\sqrt{2}, \sqrt{3}]$ and is therefore closed in $\mathbf{Q}$. However, E is *not* a compact subset of $\mathbf{Q}$. To see this, it is sufficient by 3.2 to notice that E is not a compact subset of $\mathbf{R}$. Indeed, E is not a closed subset of $\mathbf{R}$, for $\sqrt{2}$ is a cluster point of E in $\mathbf{R}$ but does not belong to E.

We shall show presently that the converse of 3.8 is valid for the metric space $\mathbf{R}^k$ with the usual metric. First, however, we shall prove a result about closed intervals in $\mathbf{R}^k$. We begin by proving the special case in which $k = 1$.

3.10 LEMMA. *Let (I_n) be a sequence of bounded, closed intervals in $\mathbf{R}$ such that $I_{n+1} \subseteq I_n$ ($n = 1, 2, \ldots$). Then $\bigcap_{n=1}^{\infty} I_n \neq \emptyset$.*

Proof. Suppose that $I_n = [a_n, b_n]$ ($n = 1, 2, \ldots$). If m, n are positive integers, then I_m and I_n both contain I_{m+n}, and so

$$a_m \leq a_{m+n} \leq b_{m+n} \leq b_n.$$

Hence, if $x = \sup\{a_n : n = 1, 2, \ldots\}$, we have $a_n \leq x \leq b_n$ for each positive integer n and $x \in \bigcap_{n=1}^{\infty} I_n$.

3.11 DEFINITION. A subset I of $\mathbf{R}^k$ is called a **closed interval** if there are k closed intervals $I_1, I_2, \ldots, I_k$ in $\mathbf{R}$ such that $I = I_1 \times I_2 \times \cdots \times I_k$, the Cartesian product of $I_1, I_2, \ldots, I_k$.

Thus I is a bounded, closed interval in $\mathbf{R}^k$ if and only if there are real numbers a_i, b_i ($a_i \leq b_i$), $i = 1, 2, \ldots, k$, such that

$$I = \{(x_1, x_2, \ldots, x_k) \in \mathbf{R}^k : a_i \leq x_i \leq b_i \text{ for } i = 1, 2, \ldots, k\}.$$

In particular, I is a bounded, closed interval in $\mathbf{R}^2$ if and only if there are real numbers a_1, b_1, a_2, b_2 ($a_1 \leq b_1, a_2 \leq b_2$) such that

$$I = \{(x_1, x_2) \in \mathbf{R}^2 : a_1 \leq x_1 \leq b_1 \text{ and } a_2 \leq x_2 \leq b_2\},$$

so that a bounded, closed interval in $\mathbf{R}^2$ is simply a closed rectangle. (See Figure 3.3.)

3.12 NESTED INTERVALS THEOREM. *Let (I_n) be a sequence of bounded, closed intervals in $\mathbf{R}^k$ such that $I_{n+1} \subseteq I_n$ ($n = 1, 2, \ldots$). Then $\bigcap_{n=1}^{\infty} I_n \neq \emptyset$.*

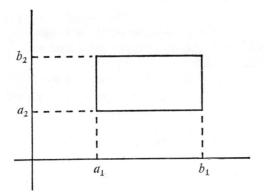

Figure 3.3

Proof. Suppose, that for each positive integer n,

$$I_n = I_{n1} \times I_{n2} \times \ldots \times I_{nk}$$

where $I_{n1}, I_{n2}, \ldots, I_{nk}$ are bounded, closed intervals in **R**. Then, for $i = 1, 2, \ldots, k$, we have $I_{(n+1)i} \subseteq I_{ni}$ ($n = 1, 2, \ldots$), and so by 3.10 there is a real number x_i in $\bigcap_{n=1}^{\infty} I_{ni}$. The point $(x_1, x_2, \ldots, x_k)$ belongs to $\bigcap_{n=1}^{\infty} I_n$.

We can now give the promised proof of the converse of 3.8 for the metric space $\mathbf{R}^k$.

3.13 HEINE-BOREL THEOREM. *Let E be a bounded, closed subset of $\mathbf{R}^k$. Then E is compact.*

Proof. Since E is bounded, it is contained in some bounded closed interval

$$I_1 = \{(x_1, x_2, \ldots, x_k) \in \mathbf{R}^k : -M \leqslant x_i \leqslant M, i = 1, 2, \ldots, k\}.$$

To show that E is compact, it is sufficient by 3.9 to show that I_1 is compact. We argue by contradiction. Suppose that there is an open covering $\mathscr{G}$ of I_1 which contains no finite subcovering of I_1. We may express I_1 as the union of 2^k closed subintervals of the form $J_1 \times J_2 \times \ldots \times J_k$ where, for $i = 1, 2, \ldots, k$, the interval J_i is either $[-M, 0]$ or $[0, M]$; these 2^k intervals are obtained by bisecting the "sides" of I_1. Since $\mathscr{G}$ contains no finite subcovering of I_1, at least one of these 2^k closed intervals cannot be covered by a finite number

of sets from $\mathscr{G}$. Let I_2 be such an interval. By bisecting the sides of I_2 and continuing this process, we obtain a sequence (I_n) of closed intervals in $\mathbf{R}^k$ such that, for each positive integer n,
(a) $I_{n+1} \subseteq I_n$,
(b) I_n cannot be covered by a finite number of sets from $\mathscr{G}$,
(c) if $x = (x_1, x_2, \ldots, x_k)$ and $y = (y_1, y_2, \ldots, y_k)$ belong to I_n, then

$$d(x, y) = \left\{ \sum_{i=1}^{k} (x_i - y_i)^2 \right\}^{\frac{1}{2}} \leq \frac{M\sqrt{k}}{2^{n-2}}.$$

Figure 3.4 illustrates the set-up when $k = 2$.

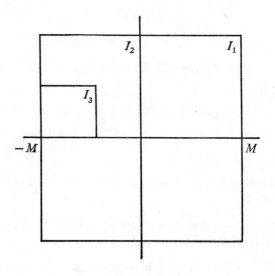

Figure 3.4

By (a) and the Nested Intervals Theorem, there is a point x in $\bigcap_{n=1}^{\infty} I_n$. Since $\mathscr{G}$ covers I_1, there is a set G in $\mathscr{G}$ such that $x \in G$. Since G is open, there is a positive real number r such that $B(x; r) \subseteq G$. If n is so large that $\dfrac{M\sqrt{k}}{2^{n-2}} < r$ (such an n certainly exists), then, by (c), $I_n \subseteq G$. Thus the singleton $\{G\}$ covers I_n, contradicting (b). The result follows.

We note that a bounded, closed subset of $\mathbf{C}$ is compact. (The proof

METRIC SPACES 17

is the same as that of 3.13 when $k = 2$.) We also refer to this result as the Heine-Borel theorem.

We have obtained a simple characterization of compact subsets of $\mathbf{R}^k$ and of $\mathbf{C}$: a subset of $\mathbf{R}^k$ or of $\mathbf{C}$ is compact if and only if it is closed and bounded. There is no such simple characterization of compact subsets of a general metric space. Indeed the description of the compact subsets of a particular metric space is often a hard, but important, theorem in analysis.

§4. Connectedness

We introduce one final topological concept, namely that of connectedness. Although we shall make limited use of this concept, it plays an important role in other branches of mathematics, particularly in complex variable theory.

4.1 DEFINITION. A subset D of a metric space X is said to be **disconnected** if there are open subsets U and V of X such that $U \cap D$ and $V \cap D$ are non-empty disjoint sets having union D. A subset of X is said to be **connected** if it is not disconnected.

It is easy to prove the analogue of 3.2 for connected sets. Thus connectedness is an intrinsic property of a set and its metric.

4.2 *Examples*. (a) The set $\mathbf{N}$ of natural numbers is a disconnected subset of $\mathbf{R}$.
$U = (-\infty, \frac{3}{2})$ and $V = (\frac{3}{2}, \infty)$ are open subsets of $\mathbf{R}$, and $U \cap \mathbf{N}$ and $V \cap \mathbf{N}$ are non-empty disjoint sets with union $\mathbf{N}$.

(b) The set $\mathbf{Q}$ of rational numbers is a disconnected subset of $\mathbf{R}$.
$U = (-\infty, \sqrt{2})$ and $V = (\sqrt{2}, \infty)$ are open subsets of $\mathbf{R}$, and $U \cap \mathbf{Q}$ and $V \cap \mathbf{Q}$ are non-empty disjoint sets with union $\mathbf{Q}$.

(c) If X is any metric space and x is a point of X, then the singleton $\{x\}$ is a connected subset of X.

We have so far given no non-trivial examples of connected sets. We now give a complete description of the connected subsets of $\mathbf{R}$.

4.3 THEOREM. *A non-empty subset E of R is connected if and only if it is an interval.*

Proof. Suppose first that E is connected. Let a, b be points of E ($a < b$) and let c be a real number such that $a < c < b$. Then $c \in E$: for otherwise $U = (-\infty, c)$ and $V = (c, \infty)$ would be open subsets of **R** such that $U \cap E$ and $V \cap E$ are non-empty disjoint sets with union E, and E would be disconnected. It is easily deduced that E is an interval having end-points inf E and sup E.

To prove the converse, we argue by contradiction. Suppose that E is an interval, but is disconnected. Then there are open subsets U, V of **R** such that $U \cap E$, $V \cap E$ are non-empty disjoint sets with union E. Choose $u \in U \cap E$ and $v \in V \cap E$. Then $u \neq v$: suppose for the sake of definiteness that $u < v$. Write $w = \sup\{x \in U : x < v\}$. Clearly $u \leqslant w \leqslant v$. Since U is open, $u \in U$ and $u < v$, there is a $\delta > 0$ such that the interval $(u-\delta, u+\delta)$ is contained in $U \cap (-\infty, v)$, and so $w > u$. Since $u, v \in E$ and E is interval, $w \in E \subseteq U \cup V$. If $w \in U$, then $w < v$ and, as above, there are numbers in U greater than w, contrary to the definition of w. If $w \in V$, then, since V is open and $w > u$, there is an $\varepsilon > 0$ such that $(w-\varepsilon, w] \subseteq V \cap E$, which implies that $(w-\varepsilon, w] \cap U$ is empty and again contradicts the definition of w. The converse follows.

4.4 COROLLARY. *The metric space* $\mathbf{R}^k$ *is connected.*

Proof. Suppose that this is not the case. Then there are non-empty disjoint open subsets U, V of $\mathbf{R}^k$ having union $\mathbf{R}^k$. Let $u \in U$, $v \in V$ and write

$$U_1 = \{t \in \mathbf{R} : u+t(v-u) \in U\}, \quad V_1 = \{t \in \mathbf{R} : u+t(v-u) \in V\}.$$

Let $t_0 \in U_1$. Then $u+t_0(v-u) \in U$ and, since U is open in $\mathbf{R}^k$, there is an $\varepsilon > 0$ such that $\{x \in \mathbf{R}^k : \| x-u-t_0(v-u) \| < \varepsilon\} \subseteq U$. If $t \in \mathbf{R}$ and $|t-t_0| < \dfrac{\varepsilon}{\|v-u\|}$, then

$$\| u+t(v-u)-u-t_0(v-u) \| = |t-t_0| \, \|v-u\| < \varepsilon,$$

so that $u+t(v-u) \in U$, i.e. $t \in U_1$. It follows that U_1 is open in **R**. Similarly V_1 is open in **R**. Since U, V are disjoint and have union $\mathbf{R}^k$, the sets U_1, V_1 are disjoint and have union **R**. Further U_1, V_1 are non-empty: $0 \in U_1$ and $1 \in V_1$. These statements imply that **R** is disconnected, the required contradiction.

4.5 COROLLARY. *The only subsets of* $\mathbf{R}^k$ *which are both open and closed are* $\emptyset$ *and* $\mathbf{R}^k$.

Proof. If U is a subset of $\mathbf{R}^k$ which is both open and closed, then U and $C(U)$ are disjoint open subsets of $\mathbf{R}^k$ with union $\mathbf{R}^k$. Since $\mathbf{R}^k$ is connected, one of these sets must be empty.

METRIC SPACES

We shall use 4.3 to obtain a simple description of the open subsets of **R**. First we introduce the idea of a component of a metric space.

4.6 DEFINITION. A subset S of a metric space X is called a **component** of X if S is a maximal connected subset of X, i.e. S is connected and if T is a connected subset of X containing S, then $T = S$.

We shall show that the components of a metric space X form a partition of X. We require:

4.7 THEOREM. *Let $\{C_i : i \in I\}$ be a collection of connected subsets of a metric space X such that $\bigcap_{i \in I} C_i$ is non-empty. Then $E = \bigcup_{i \in I} C_i$ is also a connected subset of X.*

Proof. Suppose that E is disconnected, and let U, V be open subsets of X such that $U \cap E$, $V \cap E$ are non-empty disjoint sets with union E. Choose $x \in \bigcap_{i \in I} C_i$. Then $x \in E \subseteq U \cup V$, say $x \in U$. Choose $y \in V \cap E$. Then there is an index i_0 in I such that $y \in C_{i_0}$. Since $x \in U \cap C_{i_0}$, $y \in V \cap C_{i_0}$, the sets $U \cap C_{i_0}$, $V \cap C_{i_0}$ are non-empty. As these sets are disjoint and U, V are open in X, these statements imply that C_{i_0} is disconnected, the required contradiction.

4.8 COROLLARY. *The distinct components of a metric space X are disjoint and have union X.*

Proof. If S, T are components of X and $S \cap T \neq \emptyset$, then by 4.7 $S \cup T$ is a connected subset of X. Since $S \cup T$ contains S and T, the maximality of components implies that $S = S \cup T = T$. Thus distinct components of X are disjoint.

If x is any point of X, then x belongs to a component of X, namely the union of the collection of connected subsets of X which contain x: since the singleton $\{x\}$ is a connected subset of X, this collection is non-empty. Hence the union of the components of X is all of X.

4.9 COROLLARY. *Let G be an open subset of* **R**. *Then G is a countable union of disjoint open intervals.*

Proof. According to 4.8, G is the union of its components. Further the components of G, being connected subsets of **R**, are intervals. We now show that these intervals are necessarily open. Let C be a component of G and $x_0 \in C$. Then $x_0 \in G$ and so, since G is open in

R, there is an $\varepsilon > 0$ such that $(x_0-\varepsilon, x_0+\varepsilon) \subseteq G$. Since the interval $(x_0-\varepsilon, x_0+\varepsilon)$ is connected, 4.7 implies that $C \cup (x_0-\varepsilon, x_0+\varepsilon)$ is a connected subset of G. As C is a component of G, we must have $C \cup (x_0-\varepsilon, x_0+\varepsilon) = C$, i.e. $(x_0-\varepsilon, x_0+\varepsilon) \subseteq C$. It follows that C is open.

It only remains to observe that G has a countable number of components. This follows since the components of G are open intervals, every open interval contains a rational number and the set of rational numbers is countable.

4.3 shows that the connected subsets of **R** are precisely those suggested by our intuition, namely those which are "all in one piece". We now obtain another boost for our intuition by establishing a useful characterization of the open connected subsets of $\mathbf{R}^k$.

4.10 DEFINITION. If $x, y \in \mathbf{R}^k$, then the set of points

$$[x, y] = \{tx+(1-t)y : 0 \leqslant t \leqslant 1\}$$

is called the **line segment** joining x and y. If $x_1, x_2, \ldots, x_n$ are n points of $\mathbf{R}^k$, then $\bigcup_{i=1}^{n-1} [x_i, x_{i+1}]$ is called a **polygonal curve** joining x_1 and x_n. A subset E of $\mathbf{R}^k$ is said to be **polygonally connected** if each pair of points in E can be joined by a polygonal curve entirely contained in E. (See Figure 4.1.)

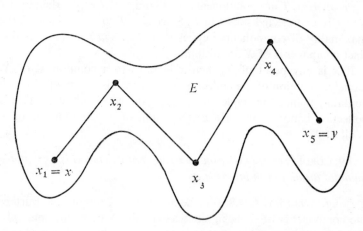

Figure 4.1. $\bigcup_{i=1}^{4} [x_i, x_{i+1}]$ is a polygonal curve joining x and y.

4.11 THEOREM. *An open subset G of $\mathbf{R}^k$ is connected if and only if it is polygonally connected.*

[A similar result holds for open subsets of $\mathbf{C}$: indeed, a similar result holds for open subsets of any normed space in which the notion of line segment is defined.]

Proof. Suppose first that G is polygonally connected but disconnected. Then there are open subsets U, V of $\mathbf{R}^k$ such that $U \cap G$, $V \cap G$ are non-empty disjoint sets having union G. Choose $u \in U \cap G$ and $v \in V \cap G$. Since G is, by hypothesis, polygonally connected, there is a finite set of points $x_1, x_2, \ldots, x_n$ in G such that $x_1 = u$, $x_n = v$ and $\bigcup_{i=1}^{n-1} [x_i, x_{i+1}] \subseteq G$. Let j be the smallest positive integer such that $x_j \in U$ and $x_{j+1} \in V$. Then, as in the proof of 4.4,

$$U_1 = \{t \in \mathbf{R}: x_j + t(x_{j+1} - x_j) \in U\},$$
$$V_1 = \{t \in \mathbf{R}: x_j + t(x_{j+1} - x_j) \in V\}$$

are open subsets of $\mathbf{R}$. The sets $U_1 \cap [0, 1]$, $V_1 \cap [0, 1]$ are non-empty and, since $[x_j, x_{j+1}] \subseteq G \subseteq U \cup V$, are disjoint and have union $[0, 1]$. These statements imply that $[0, 1]$ is disconnected, a contradiction.

Conversely, suppose that G is connected. Let x be a point of G,

$$U = \{y \in G: \text{there is a polygonal curve in } G \text{ joining } x \text{ and } y\}$$

and

$$V = \{y \in G: \text{there is no polygonal curve in } G \text{ joining } x \text{ and } y\}.$$

$x \in U$ and so U is non-empty. Let $y_0 \in U$. Then, since G is open, there is an $\varepsilon > 0$ such that $B(y_0; \varepsilon) \subseteq G$. If $y \in B(y_0; \varepsilon)$, then $[y_0, y] \subseteq B(y_0; \varepsilon)$. It follows that $B(y_0; \varepsilon) \subseteq U$. (See Figure 4.2.)

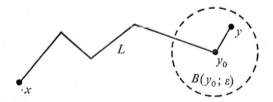

Figure 4.2. L is a polygonal curve in G joining x and y_0; $L \cup [y_0, y]$ is a polygonal curve in G joining x and y.

Hence U is open. A similar argument shows that V is open. U and V are disjoint sets and have union G. Since G is connected, V cannot be non-empty. Hence $V = \emptyset$ and $U = G$, i.e. G is polygonally connected.

The first part of the above proof shows that if E is a polygonally connected subset of $\mathbf{R}^k$, then E is connected. It must, however, be emphasized that the converse is not true unless E is assumed to be open.

§5. Convergence

5.1 DEFINITION. A sequence (x_n) in a metric space (X, d) is said to **converge to a point** x of X if for each $\varepsilon > 0$, there is a positive integer n_0 such that $d(x_n, x) < \varepsilon$ for all $n \geqslant n_0$. The sequence (x_n) is said to **converge** (in X) if there is a point in X to which it converges.

5.1 is an obvious generalization of the classical definition of the convergence of a sequence of real or complex numbers. It is, however, worth pointing out that we could well have defined convergence in terms of neighbourhoods.

5.2 THEOREM. *Let (x_n) be a sequence in a metric space X and x be a point of X. Then (x_n) converges to x if and only if, for each neighbourhood V of x, there is a positive integer n_0 such that $x_n \in V$ for all $n \geqslant n_0$.*

Proof. Suppose that (x_n) converges to x, and let V be a neighbourhood of x. Then there is an $\varepsilon > 0$ such that $B(x; \varepsilon) \subseteq V$. By hypothesis, there is a positive integer n_0 such that $d(x_n, x) < \varepsilon$ for all $n \geqslant n_0$. Then $x_n \in V$ for all $n \geqslant n_0$, and the property of the theorem holds.

Conversely, suppose that the property of the theorem holds, and let $\varepsilon > 0$. Then $B(x; \varepsilon)$ is a neighbourhood of x, and there is a positive integer n_0 such that $x_n \in B(x; \varepsilon)$, i.e. $d(x_n, x) < \varepsilon$, for all $n \geqslant n_0$.

It is easily shown that a sequence in a metric space X cannot converge to distinct points of X. Hence, if a sequence (x_n) in X converges to x, we may refer to x as *the* limit of the sequence (x_n) and write $x = \lim_{n \to \infty} x_n$ or simply $x = \lim x_n$.

At the end of §3, we pointed out that it is frequently difficult to prove that a subset of a metric space is compact. It is now possible to give a useful characterization of compactness.

5.3 DEFINITION. A subset K of a metric space X is said to be **sequentially compact** if every sequence in K has a subsequence which converges to a point of K.

5.4 THEOREM. *Let K be a subset of a metric space X. Then K is compact if and only if it is sequentially compact.*

Proof. Suppose first that K is compact, but is not sequentially compact. Then there is a sequence (x_n) in K which has no subsequence convergent to a point of K. The set $S = \{x_1, x_2, x_3, \ldots\}$ can have no cluster point in K: for if x belongs to K and is a cluster point of S, we could choose inductively a strictly increasing sequence (k_n) of positive integers such that $d(x_{k_n}, x) < n^{-1}$ $(n = 1, 2, \ldots)$, so that the subsequence (x_{k_n}) of (x_n) converges to x. Hence each point x of K has an open neighbourhood V_x which contains at most one point of S (namely x, if $x \in S$). $\{V_x : x \in K\}$ is an open covering of K, and so contains a finite subcovering of K. Thus the set S is finite, and the sequence (x_n) most certainly does have a subsequence which converges to a point of K. This contradiction shows that K is sequentially compact.

The proof of the converse is lengthier, and is left to the problems at the end of the chapter.

According to 5.1, to show that a sequence (x_n) in a metric space X converges, one must find a point x in X such that (x_n) converges to x, i.e. one must guess the limit of the sequence and then justify the guess. The General Principle of Convergence, one of the most important results in an introductory analysis course, gives a criterion for the convergence of a sequence of real or complex numbers which does not involve guessing the limit of the sequence. We now discuss the analogous result for sequences in a general metric space.

5.5 DEFINITION. A sequence (x_n) in a metric space X is said to be **Cauchy** (or **fundamental**) if for each $\varepsilon > 0$, there is a positive integer n_0 such that $d(x_p, x_q) < \varepsilon$ for all $p, q \geqslant n_0$.

5.6 THEOREM. *Every convergent sequence is Cauchy.*

Proof. Let (x_n) be a convergent sequence, with limit x say, and ε be a positive real number. Then there is a positive integer n_0 such that

$$d(x_n, x) < \tfrac{1}{2}\varepsilon \text{ for all } n \geqslant n_0.$$

Hence, if $p, q \geqslant n_0$,
$$d(x_p, x_q) \leqslant d(x_p, x) + d(x, x_q) < \tfrac{1}{2}\varepsilon + \tfrac{1}{2}\varepsilon = \varepsilon.$$
Thus the sequence (x_n) is Cauchy.

The converse of this result is false in general. For example, the sequence $((1+n^{-1})^n)$ of rational numbers is Cauchy but does not converge in **Q**. (Remember $\lim (1+n^{-1})^n = e$, which is irrational.) In view of the importance of the General Principle of Convergence, it is natural to single out those metric spaces for which the converse of 5.6 is valid.

5.7 DEFINITION. A metric space X is said to be **complete** if every Cauchy sequence in X converges in X.

We now examine the completeness of some of the metric spaces introduced in §1. By the General Principle of Convergence, **R** and **C** are both complete.

To see that $\mathbf{R}^k$ is complete, let (x_n) be a Cauchy sequence in $\mathbf{R}^k$. Suppose that $x_n = (x_{n1}, x_{n2}, \ldots, x_{nk})$, $n = 1, 2, \ldots$. Then, for $j = 1, \ldots, k$,
$$|x_{pj} - x_{qj}| \leqslant \left\{ \sum_{i=1}^{k} (x_{pi} - x_{qi})^2 \right\}^{\frac{1}{2}} = d(x_p, x_q)$$
p and q being any positive integers. Thus the k sequences (x_{nj}) are Cauchy in **R**. Since **R** is complete, there are real numbers $y_1, \ldots, y_k$ such that
$$y_j = \lim_{n \to \infty} x_{nj} \quad (j = 1, \ldots, k).$$
If $y = (y_1, y_2, \ldots, y_k)$, then
$$\lim_{n \to \infty} d(x_n, y) = \lim_{n \to \infty} \left\{ \sum_{i=1}^{k} (x_{ni} - y_i)^2 \right\}^{\frac{1}{2}} = 0,$$
so that the sequence (x_n) converges to y.

The metric space **m** of all bounded sequences of complex numbers is also complete. For let (x_n) be a Cauchy sequence in **m**, and suppose that the ith term of x_n is x_{ni}. Then, for each positive integer i,
$$|x_{pi} - x_{qi}| \leqslant \sup_{j \geqslant 1} |x_{pj} - x_{qj}| = d(x_p, x_q).$$
Thus, for each positive integer i, the sequence (x_{ni}) of complex

numbers is Cauchy, and there is a complex number y_i such that $\lim_{n \to \infty} x_{ni} = y_i$. Denote the sequence having ith term y_i by y. We show that $y \in \mathbf{m}$ and that (x_n) converges to y in $\mathbf{m}$. Since the sequence x_n of complex numbers is bounded, there is a number K_n such that

$$|x_{ni}| \leq K_n \text{ for all } i \quad (n = 1, 2, \ldots).$$

Let $\varepsilon > 0$. Since the sequence (x_n) of elements of $\mathbf{m}$ is Cauchy, there is a positive integer n_0 such that

$$d(x_p, x_q) = \sup_{i \geq 1} |x_{pi} - x_{qi}| < \varepsilon \text{ for all } p, q \geq n_0. \quad (5.1)$$

Using (5.1), we have, for each positive integer i,

$$|y_i - x_{qi}| = \lim_{p \to \infty} |x_{pi} - x_{qi}| \leq \varepsilon \text{ for all } q \geq n_0.$$

Thus $|y_i| \leq |y_i - x_{n_0 i}| + |x_{n_0 i}| \leq \varepsilon + K_{n_0} \; (i = 1, 2, \ldots)$,

i.e. (y_n) is a bounded sequence of complex numbers and therefore belongs to $\mathbf{m}$. Further, for all $q \geq n_0$,

$$d(y, x_q) = \sup_{i \geq 1} |y_i - x_{qi}| \leq \varepsilon,$$

so that $\lim_{n \to \infty} x_n = y$.

$C[a, b]$ with metric d given by $d(x, y) = \int_a^b |x(t) - y(t)| \, dt$ is *not* complete. To see this, let $c \in (a, b)$ and, for each positive integer n, let x_n be the function in $C[a, b]$ defined by

$$x_n(t) = \begin{cases} 0 & (a \leq t \leq c - 1/n) \\ n(t - c + 1/n) & (c - 1/n < t \leq c) \\ 1 & (c < t \leq b) \end{cases}$$

(See Figure 5.1.)

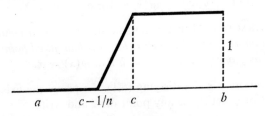

Figure 5.1. The graph of x_n.

If p, q are positive integers with $q \geq p$, then

$$d(x_p, x_q) = \int_a^b |x_p(t) - x_q(t)| \, dt = \tfrac{1}{2}p - \tfrac{1}{2}q,$$

so that the sequence (x_n) is Cauchy. However, if $x \in C[a, b]$, then

$$d(x_n, x) = \int_a^b |x_n(t) - x(t)| \, dt$$
$$= \int_a^{c-1/n} |x(t)| \, dt + \int_{c-1/n}^c |x_n(t) - x(t)| \, dt + \int_c^b |1 - x(t)| \, dt.$$

Hence $\lim_{n \to \infty} d(x_n, x) = 0$ implies that

$$\int_a^c |x(t)| \, dt = 0 = \int_c^b |1 - x(t)| \, dt,$$

which in turn imply that

$$x(t) = 0 \text{ for } a \leq t \leq c \quad \text{and} \quad x(t) = 1 \text{ for } c \leq t \leq b.$$

Since this is ridiculous, the sequence (x_n) does not converge in $(C[a, b], d)$.

We shall discuss the completeness of $C[a, b]$, with the uniform metric, in a later section.

§6. Consequences of completeness

In this section, we shall prove two results which depend essentially on the concept of completeness. The first result, about contractions, provides a useful tool for the proof of many existence and uniqueness theorems.

6.1 DEFINITION. A function φ of a metric space X into itself is called a **contraction** on X if there is a number $\alpha < 1$ such that

$$d(\varphi(x), \varphi(y)) \leq \alpha d(x, y) \text{ for all } x, y \text{ in } X. \tag{6.1}$$

6.2 CONTRACTION MAPPING THEOREM. *Let φ be a contraction on a complete metric space X. Then φ has a unique fixed point, i.e. there is one and only one point x in X such that $\varphi(x) = x$.*

Proof. Choose $\alpha < 1$ satisfying (6.1). To prove the existence of a fixed point of φ, let x_1 be any point of X and write

$$x_{n+1} = \varphi(x_n) \quad (n = 1, 2, \ldots).$$

METRIC SPACES 27

Then, for each positive integer n,

$$d(x_{n+1}, x_{n+2}) = d(\varphi(x_n), \varphi(x_{n+1}))$$
$$\leqslant \alpha d(x_n, x_{n+1}) \leqslant \alpha^2 d(x_{n-1}, x_n) \leqslant \ldots \leqslant \alpha^n d(x_1, x_2).$$

Hence, if p, q are positive integers with $q > p$, we have

$$d(x_p, x_q) \leqslant d(x_p, x_{p+1}) + d(x_{p+1}, x_{p+2}) + \ldots + d(x_{q-1}, x_q)$$
$$\leqslant [\alpha^{p-1} + \alpha^p + \ldots + \alpha^{q-2}] d(x_1, x_2)$$
$$\leqslant \alpha^{p-1}[1 + \alpha + \alpha^2 + \ldots] d(x_1, x_2)$$
$$= \frac{\alpha^{p-1}}{1-\alpha} d(x_1, x_2).$$

It follows that the sequence (x_n) is Cauchy, and therefore, since X is complete, converges to a point x of X. Since

$$d(\varphi(x), x_{n+1}) = d(\varphi(x), \varphi(x_n)) \leqslant \alpha d(x, x_n),$$

$\varphi(x) = \lim x_{n+1} = x$. Thus x is a fixed point of φ.

If y is any fixed point of φ, then

$$d(x, y) = d(\varphi(x), \varphi(y)) \leqslant \alpha d(x, y),$$

which implies that $d(x, y) = 0$, i.e. $x = y$.

6.3 BAIRE'S THEOREM. *Let (F_n) be a sequence of closed subsets of a complete metric space X such that $\bigcup_{n=1}^{\infty} F_n = X$. Then at least one of the sets F_n contains a non-empty open set.*

Proof. We argue by contradiction. Suppose that none of the sets F_n contains a non-empty open set. Then $F_1 \neq X$, and $X - F_1$ is open and non-empty. Choose $x_1 \in X - F_1$ and $r_1 \in (0, 1)$ so that $B(x_1; r_1) \subseteq X - F_1$: $B(x_1; r_1) \cap F_1 = \emptyset$. By hypothesis, $B(x_1; \frac{1}{2}r_1)$ is not contained in F_2. Since F_2 is closed, we can choose x_2 in $B(x_1; \frac{1}{2}r_1)$ and $r_2 > 0$ such that $B(x_2; r_2) \cap F_2 = \emptyset$: we may assume that r_2 is chosen so that $r_2 < \frac{1}{2}$ and $B(x_2; r_2) \subseteq B(x_1; \frac{1}{2}r_1)$ also. Proceeding inductively, we may choose a sequence (x_n) of points in X and a sequence (r_n) of positive real numbers such that $r_n < n^{-1}$, $B(x_{n+1}; r_{n+1}) \subseteq B(x_n; \frac{1}{2}r_n)$ and $B(x_n; r_n) \cap F_n = \emptyset$ for $n = 1, 2, \ldots$. If p, q are positive integers with $q > p$, then

$$B(x_q; r_q) \subseteq B(x_{q-1}; \tfrac{1}{2}r_{q-1})$$
$$\subseteq B(x_{q-1}; r_{q-1}) \subseteq \ldots \subseteq B(x_{p+1}; r_{p+1}) \subseteq B(x_p; \tfrac{1}{2}r_p),$$

so that $d(x_p, x_q) < \frac{1}{2}r_p$. Since $\lim_{p \to \infty} r_p = 0$, it follows that the sequence (x_n) is Cauchy, and hence, since X is complete, converges to a point x of X. Since

$$d(x, x_p) \leq d(x, x_q) + d(x_q, x_p)$$
$$< d(x, x_q) + \frac{1}{2}r_p \quad (q > p)$$
$$\to \frac{1}{2}r_p \quad \text{as} \quad q \to \infty,$$

$x \in B(x_p; r_p)$ which implies that $x \notin F_p$ ($p = 1, 2, \ldots$). As $\bigcup_{n=1}^{\infty} F_n = X$ this gives the required contradiction.

Although Baire's theorem is of considerable importance in functional analysis, its significance is rather difficult to appreciate at this stage. We note that the completeness hypothesis is essential. For the metric space **Q** is a countable union of singletons, and singletons are closed sets which contain no non-empty open sets.

There is some rather undescriptive terminology often used in connection with Baire's theorem.

6.4 DEFINITION. A subset E of a metric space is said to be of the **first category** if it can be expressed as a countable union of **nowhere dense** sets, i.e. sets whose closures do not contain any non-empty open sets; otherwise E is said to be of the **second category**.

The above theorem, sometimes called Baire's category theorem, may be restated as follows: *every complete metric (considered as a subset of itself) is of the second category.*

Problems on Chapter 1

1. Let X be a non-empty set and define the function d on $X \times X$ by

$$d(x, y) = \begin{cases} 1 & \text{if } x \neq y \\ 0 & \text{if } x = y. \end{cases}$$

Show that d is a metric on X. (d is called the **discrete** metric on X.)

2. Show that l^2, the set of all sequences (x_n) of complex numbers such that $\sum_{n=1}^{\infty} |x_n^2| < \infty$, is a vector space under the usual algebraic operations and that

$$\|x\| = \left\{ \sum_{n=1}^{\infty} |x_n^2| \right\}^{\frac{1}{2}} \quad (x = (x_n) \in l^2)$$

defines a norm on l^2.

3. Let d^1 be the function on $\mathbf{R}^k \times \mathbf{R}^k$ defined by

$$d^1(x, y) = \max_{1 \leq i \leq k} |x_i - y_i| \quad (x = (x_1, \ldots, x_k), \ y = (y_1, \ldots, y_k)),$$

and denote the usual metric on $\mathbf{R}^k$ by d. Show that
(a) d^1 is a metric on $\mathbf{R}^k$;
(b) there are positive numbers a, b such that

$$ad^1(x, y) \leq d(x, y) \leq bd^1(x, y) \text{ for all } x, y \in \mathbf{R}^k;$$

(c) the metrics d and d^1 are **equivalent**, i.e. the metric spaces $(\mathbf{R}^k, d)$ and $(\mathbf{R}^k, d^1)$ have the same open subsets.

4. Let A be a subset of a metric space X and x be a cluster point of A. Show that every neighbourhood of x contains infinitely many points of A.

5. What are the cluster points of $\{(m/n, 1/n) : m, n \text{ positive integers}\}$ in $\mathbf{R}^2$?

6. If d is the discrete metric on X, which subsets of (X, d) are (a) open, (b) compact?

7. According to 3.8, $(0, 1)$ is not a compact subset of $\mathbf{R}$. Produce an open covering of $(0, 1)$ which contains no finite subcovering.

8. According to 3.2 and 3.8, $\{x \in \mathbf{Q} : 0 \leq x \leq 1\}$ is not a compact metric space. Produce an open covering of this space which contains no finite subcovering.

9. Let X be a metric space, F be a closed subset of X and K be a compact subset of X such that $F \cap K = \emptyset$. Show that
(a) for each $x \in K$, there is a positive real number r_x and an open neighbourhood U_x of x such that

$$d(u, y) \geq r_x \text{ for all } u \in U_x \text{ and } y \in F;$$

(b) $\inf\{d(x, y) : x \in K, y \in F\} > 0$.
Show that (b) need not be true if K is only assumed to be closed.

10. Let U be an open subset of $\mathbf{R}^k$ and C be a compact subset of $\mathbf{R}^k$ such that $C \subseteq U$. Show that there is a compact subset D of $\mathbf{R}^k$ such that $C \subseteq D^0$ and $D \subseteq U$.

11. Show that a subset F of a metric space X is closed if and only if the limit of each convergent sequence of points of F belongs to F.

12. Let K be a sequentially compact subset of a metric space X. Show that
   (a) if $\delta > 0$, the open covering $\{B(x; \delta): x \in K\}$ of K contains a finite subcovering $\mathcal{G}_\delta$ of K;
   (b) if $x \in K$ and V is an open neighbourhood of x, there is a positive integer n and a ball B in $\mathcal{G}_{1/n}$ such that $x \in B \subseteq V$;
   (c) if $\{V_i: i \in I\}$ is an open covering of K, there is a countable subset J of I such that $\bigcup_{j \in J} V_j \supseteq K$;

   [Hint: let $\mathcal{G}$ be the subfamily of $\bigcup_{n=1}^{\infty} \mathcal{G}_{1/n}$ consisting of those balls B which are contained in some V_i: $\mathcal{G}$ is countable. For each $B \in \mathcal{G}$, choose one set V_i containing B. If $\{V_j: j \in J\}$ denotes the collection of V_i's chosen, then $\bigcup_{j \in J} V_j \supseteq K$.]
   (d) K is compact.
   [Hint: (c) shows that it is sufficient to prove that every countable open covering of K contains a finite subcovering.]

   Note: the set of centres of the balls in the families $\mathcal{G}_{1/n}, n = 1, 2, \ldots,$ is countable and is **dense** in K, i.e. its closure contains K. We shall use the fact that a compact metric space has a countable dense subset later on.

13. Show that the subset $\{(x_n) \in \mathbf{m}: \sup_{n \geq 1} |x_n| \leq 1\}$ of $\mathbf{m}$ is not compact.

14. Let A be a connected subset of a metric space X and B be a subset of X such that $A \subseteq B \subseteq A^-$. Show that B is connected.

15. Show that an open subset of $\mathbf{R}^k$ can be expressed as a countable union of open balls.

16. Let A be a subset of a metric space X, and define the **boundary** of A, ∂A, by $\partial A = A^- \cap (X - A)^-$. Show that
   (a) $x \in \partial A$ if and only if every neighbourhood of x contains a point of A and a point of $X - A$;
   (b) A is closed if and only if $\partial A \subseteq A$;
   (c) A is open if and only if $\partial A \cap A = \varnothing$.

17. Show that the metric space **s** is complete.

18. Let X be a complete metric space and A be a subset of X. Show

that A is closed in X if and only if A, with the relative metric, is a complete metric space.

19. Show that **c**, the set of all convergent sequences of complex numbers, is a closed subset of **m**.

20. Let X be a complete metric space, x_0 an element of X, δ a positive real number and φ a function of the closed ball B in X, having centre x_0 and radius δ, into X. Suppose that there is a number α in $(0, 1)$ such that

$$d(\varphi(x), \varphi(y)) \leq \alpha d(x, y) \quad \text{for all } x, y \text{ in } B$$

and

$$d(\varphi(x_0), x_0) \leq \delta(1-\alpha).$$

Show that
(a) φ has a unique fixed point u in B;
(b) if $x \in B$ and $d(\varphi(x), x) \leq r$, then $d(x, u) \leq \dfrac{r}{1-\alpha}$.

CHAPTER 2
CONTINUOUS FUNCTIONS

§7. Definition and topological conditions

7.1 DEFINITION. Let f be a function of a metric space (X, d_X) into a metric space (Y, d_Y). f is said to be **continuous at a point** x_0 of X if for each $\varepsilon > 0$, there is a $\delta > 0$ such that $d_Y(f(x), f(x_0)) < \varepsilon$ for each x in X satisfying $d_X(x, x_0) < \delta$.

The function f is said to be **continuous (on X)** if it is continuous at each point of X.

7.1 is an obvious generalization of the familiar definition of the continuity of a real-valued function on **R**. We now show that continuity could have been defined in terms of neighbourhoods. This result is analogous to 5.2.

7.2 THEOREM. *Let f be a function of a metric space X into a metric space Y and x_0 be a point of X. Then f is continuous at x_0 if and only if, for each neighbourhood V of $f(x_0)$, there is a neighbourhood U of x_0 such that $f(U) \subseteq V$, i.e. such that $f(x) \in V$ for each $x \in U$.*

Proof. Suppose first that f is continuous at x_0, and let V be a neighbourhood of $f(x_0)$. Then there is an $\varepsilon > 0$ such that the ball $\{y \in Y : d_Y(y, f(x_0)) < \varepsilon\}$ is contained in V. Since f is continuous at x_0, there is a $\delta > 0$ such that $d_Y(f(x), f(x_0)) < \varepsilon$ for all x in X satisfying $d_X(x, x_0) < \delta$. $U = \{x \in X : d_X(x, x_0) < \delta\}$ is a neighbourhood of x_0 and $f(U) \subseteq V$. Hence the property of the theorem holds.

Conversely, suppose that the property of the theorem holds, and let $\varepsilon > 0$. $V = \{y \in Y : d_Y(y, f(x_0)) < \varepsilon\}$ is a neighbourhood of $f(x_0)$, and so, by hypothesis, there is a neighbourhood U of x_0 such that $f(U) \subseteq V$. As U is a neighbourhood of x_0, there is a $\delta > 0$ such that $\{x \in X: d_X(x, x_0) < \delta\} \subseteq U$. Then, for each $x \in X$ satisfying $d_X(x, x_0) < \delta$, we have $f(x) \in V$, i.e. $d_Y(f(x), f(x_0)) < \varepsilon$. It follows that f is continuous at x_0.

In the next section we shall be interested in properties of functions which are continuous at each point of their domain. In this context the importance of the next result will soon be evident.

7.3 THEOREM. *Let f be a function of a metric space X into a metric space Y. Then the following statements are equivalent:*
(a) *f is continuous on X;*
(b) *for each open subset G of Y, $f^{-1}(G)$ is an open subset of X;*
(c) *for each closed subset H of Y, $f^{-1}(H)$ is a closed subset of X.*

Proof. (a) $\Rightarrow$ (b). Suppose that (a) is satisfied, and let G be an open subset of Y. Let $x_0 \in f^{-1}(G) = \{x \in X: f(x) \in G\}$. Then G is a neighbourhood of $f(x_0)$. By 7.2, there is a neighbourhood U of x_0 such that $f(U) \subseteq G$, i.e. $U \subseteq f^{-1}(G)$. It follows that $f^{-1}(G)$ is a neighbourhood of each of its points and is therefore open.

(b) $\Rightarrow$ (a). Suppose that (b) is satisfied, and let $x_0 \in X$. Let V be a neighbourhood of $f(x_0)$. Then there is an open set G such that $f(x_0) \in G \subseteq V$. Write $U = f^{-1}(G)$. Then $x_0 \in U$ and U is open, so that U is a neighbourhood of x_0, and $f(U) \subseteq G \subseteq V$. It follows by 7.2 that f is continuous at x_0, an arbitrary point of X.

(b) $\Leftrightarrow$ (c). We first note that if S is any subset of Y, then

$$f^{-1}(Y-S) = \{x \in X: f(x) \in Y-S\}$$
$$= \{x \in X: f(x) \notin S\}$$
$$= \{x \in X: x \notin f^{-1}(S)\} = X - f^{-1}(S).$$

Now suppose (b) is satisfied, and let H be a closed subset of Y. Then $Y-H$ is open in Y, and so, by (b), $f^{-1}(Y-H) = X - f^{-1}(H)$ is open in X, i.e. $f^{-1}(H)$ is a closed subset of X.

The proof that (c) implies (b) is similar.

The reader should note that 7.3 does *not* assert that if f is continuous on X and G is an open subset of X, then the direct image $f(G)$ is an open subset of Y. In general, a continuous function need not send open sets into open sets. For example, consider the function f of $\mathbf{R}$ into $\mathbf{R}$ defined by $f(x) = 1/(1+x^2)$ ($x \in \mathbf{R}$) and the set $G = (-1, 1)$. f is continuous and G is open, but $f(G) = (\tfrac{1}{2}, 1]$ is not open. (See Figure 7.1.) Similarly, a continuous function need not send closed sets into closed sets.

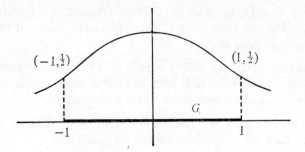

Figure 7.1 The graph of f.

It is traditional, in a first course on analysis, to introduce the notion of limit, and to define continuity in terms of limits. We have chosen to define continuity first as it will be the more important concept for us. Nevertheless it is worth giving a definition of limit, relevant to the general situation under discussion.

7.4 DEFINITION. Let X and Y be metric spaces, A a subset of X, f a function of A into Y, x_0 a cluster point of A and y a point of Y. Then we write

$$f(x) \to y \text{ as } x \to x_0 \text{ or } \lim_{x \to x_0} f(x) = y$$

if for each $\varepsilon > 0$, there is a $\delta > 0$ such that $d_Y(f(x), y) < \varepsilon$ for each x in A satisfying $0 < d_X(x, x_0) < \delta$.

The assumption that x_0 is a cluster point of A ensures that for any $\delta > 0$, there are points x in A satisfying $0 < d_X(x, x_0) < \delta$. Clearly f is continuous at x_0 if and only if $x_0 \in A$, the domain of f, and $\lim_{x \to x_0} f(x) = f(x_0)$.

§8. Preservation of compactness and connectedness

8.1 THEOREM. *Let f be a continuous function of a compact metric space X into a metric space Y. Then $f(X)$ is a compact subset of Y.*

Proof. Let $\{G_i : i \in I\}$ be an open covering of $f(X)$. By 7.3, $\{f^{-1}(G_i) : i \in I\}$ is an open covering of X. Since X is compact, there is a finite subset J of I such that $\{f^{-1}(G_i) : i \in J\}$ covers X. Then $f(X) = f(\bigcup_{i \in J} f^{-1}(G_i)) \subseteq \bigcup_{i \in J} G_i$, so that $\{G_i : i \in J\}$ covers $f(X)$. It follows that $f(X)$ is compact.

8.2 COROLLARY. *Let f be a continuous real-valued function on a compact metric space X. Then $f(X)$ is a bounded subset of $\mathbf{R}$ and there are points x_0 and x_1 in X such that*

$$f(x_0) = \sup f(X) \quad \text{and} \quad f(x_1) = \inf f(X).$$

Proof. We prove the first result: the second can be proved in a similar way. By 8.1, $f(X)$ is compact and therefore bounded and closed. Write $M = \sup f(X)$. If $\varepsilon > 0$, there is a number y in $f(X)$ such that $M \geqslant y > M - \varepsilon$. It follows that $M \in f(X)^- = f(X)$, i.e. there is a point x_0 in X such that $f(x_0) = M = \sup f(X)$.

8.3 THEOREM. *Let f be a continuous one-one function of a compact metric space X onto a metric space Y. Then f^{-1}, the inverse of f, is a continuous function of Y onto X.*

Proof. Let H be a closed subset of X. Then, since X is compact, H is compact. Hence, by 8.1, $f(H)$ is compact. Thus $(f^{-1})^{-1}(H) = f(H)$ is a closed subset of Y. The result now follows from 7.3 applied to f^{-1}.

8.2 asserts that a continuous real-valued function on a compact metric space has a largest and a smallest value, and is one of the most important results concerning compactness. To emphasize the significance of this result, we look at two applications.

8.4 FUNDAMENTAL THEOREM OF ALGEBRA. *The field of complex numbers is **algebraically closed**, i.e. every non-constant polynomial with complex coefficients has a complex zero.*

Proof. Let $a_0, a_1, \ldots, a_n$ be $(n+1)$ complex numbers, $n \geqslant 1$, with $a_n \neq 0$, and let $P(z) = a_0 + a_1 z + \ldots + a_n z^n$. Write

$$\mu = \inf \{|P(z)| : z \in \mathbf{C}\}.$$

Then we have

$$|P(z)| = |z|^n \left| a_n + \frac{a_{n-1}}{z} + \ldots + \frac{a_0}{z^n} \right| \quad (z \neq 0)$$

$$\geqslant |z|^n \left\{ |a_n| - \frac{|a_{n-1}|}{|z|} - \ldots - \frac{|a_0|}{|z|^n} \right\}$$

which is greater than μ if $|z|$ is sufficiently large. Hence there is a $\rho > 0$ such that $\mu = \inf \{|P(z)| : z \in \mathbf{C}, |z| \leqslant \rho\}$. The set

$E = \{z \in \mathbf{C} : |z| \leq \rho\}$ is compact by the Heine-Borel theorem, and $|P|$ is a continuous real-valued function on E. Hence, by 8.2, there is a point z_0 in E such that $|P(z_0)| = \mu$. We complete the proof by showing that $\mu = 0$. Suppose that this is not the case. Then

$$P(z+z_0) = a_0 + a_1(z+z_0) + \ldots + a_n(z+z_0)^n$$
$$= b_0 + b_1 z + \ldots + b_n z^n \text{ say,}$$

rearranging terms, where $b_0 = P(z_0) \neq 0$ and $b_n = a_n \neq 0$. Let k be the smallest positive integer such that $b_k \neq 0$. Then

$$|P(z+z_0)| = |b_0 + b_k z^k + b_{k+1} z^{k+1} + \ldots + b_n z^n|$$
$$\leq |b_0 + b_k z^k| + |b_{k+1}||z|^{k+1} + \ldots + |b_n||z|^n.$$

If $|z| = r$ is small and $\theta = \arg z$ is chosen so that $\arg(b_k e^{ik\theta}) = \arg(-b_0)$, then

$$|b_0 + b_k z^k| = |b_0| - |b_k z^k| \quad \text{(see Figure 8.1)}$$
$$= \mu - r^k |b_k|$$

and

$$|P(z+z_0)| \leq \mu - r^k \{|b_k| - r|b_{k+1}| - \ldots - r^{n-k}|b_n|\}. \quad (8.1)$$

If r is sufficiently small, the right-hand side of (8.1) is less than μ and (8.1) then contradicts the definition of μ.

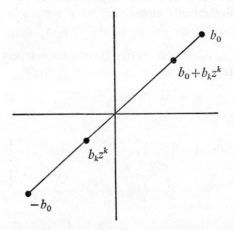

Figure 8.1. b_0 and $b_k z^k$ plotted in the complex plane.

CONTINUOUS FUNCTIONS

The second application gives an extension of the Heine-Borel theorem.

8.5 THEOREM. *Let X be a real normed space and F be a bounded, closed subset of a finite-dimensional linear subspace M of X. Then F is compact.*

Proof. Let $\{x_1, \ldots, x_k\}$ be a basis for M, and define the function T of $\mathbf{R}^k$ into M by

$$T(\lambda_1, \ldots, \lambda_k) = \sum_{i=1}^{k} \lambda_i x_i \quad ((\lambda_1, \ldots, \lambda_k) \in \mathbf{R}^k).$$

Since $\{x_1, \ldots, x_k\}$ is a basis for M, T is one-one and has range M. If $\lambda = (\lambda_1, \ldots, \lambda_k)$ and $\mu = (\mu_1, \ldots, \mu_k)$ are points of $\mathbf{R}^k$, then

$$\| T(\lambda) - T(\mu) \| = \left\| \sum_{i=1}^{k} (\lambda_i - \mu_i) x_i \right\|$$

$$\leqslant \sum_{i=1}^{k} |\lambda_i - \mu_i| \, \| x_i \|$$

$$\leqslant \left\{ \sum_{i=1}^{k} (\lambda_i - \mu_i)^2 \right\}^{\frac{1}{2}} \left\{ \sum_{i=1}^{k} \| x_i \|^2 \right\}^{\frac{1}{2}} \quad \text{using Cauchy's inequality}$$

$$= \| \lambda - \mu \| \left\{ \sum_{i=1}^{k} \| x_i \|^2 \right\}^{\frac{1}{2}},$$

where we are using the same symbol to denote the norm on X and the usual norm on $\mathbf{R}^k$. It follows that T is continuous on $\mathbf{R}^k$.

Write $K = T^{-1}(F)$. Then $F = T(K)$ and so, to prove that F is compact, it is sufficient, by 8.1, to show that K is a compact subset of $\mathbf{R}^k$. By the Heine-Borel theorem, it is enough to show that K is closed and bounded. By 7.3, K is closed. We now prove that K is bounded. Since F is bounded, there is an $m > 0$ such that $F \subseteq \{x \in X : \| x \| \leqslant m\}$. Write $\alpha = \inf_{\lambda \in S} \| T(\lambda) \|$ where $S = \{\lambda \in \mathbf{R}^k : \| \lambda \| = 1\}$. Since S is compact and the function $\lambda \to \| T(\lambda) \|$ is continuous, there is, by 8.2, a point $\lambda^0 = (\lambda_1^0, \ldots, \lambda_k^0)$ in S such that

$$\alpha = \| T(\lambda^0) \| = \left\| \sum_{i=1}^{k} \lambda_i^0 x_i \right\|.$$

As the vectors $x_1, \ldots, x_k$ are linearly independent, it follows that $\alpha > 0$. For each non-zero vector λ in $\mathbf{R}^k$,

$$\|T(\lambda)\| = \|\lambda\| \left\|T\left(\frac{\lambda}{\|\lambda\|}\right)\right\| \quad (T \text{ is linear})$$

$$\geq \alpha \|\lambda\| \quad \left(\frac{\lambda}{\|\lambda\|} \in S\right);$$

this inequality is also satisfied when $\lambda = 0$. Hence, for each $\lambda \in K$,

$$\|\lambda\| \leq \frac{1}{\alpha} \|T(\lambda)\| \leq \frac{m}{\alpha}.$$

This completes the proof.

8.6 COROLLARY. *Let X be a real normed space, M be a finite-dimensional linear subspace of X and x_0 be a point of X. Then there is a point y_0 in M such that*

$$\|x_0 - y_0\| = \inf_{y \in M} \|x_0 - y\|.$$

Proof. Let $F = \{y \in M : \|x_0 - y\| \leq \|x_0\|\}$. (See Figure 8.2.) Then F is a bounded, closed subset of M, and so, by 8.5, is compact. Further, since M contains the zero element of X, we have

$$\inf_{y \in F} \|x_0 - y\| = \inf_{y \in M} \|x_0 - y\|.$$

The nearest point theorem, 3.6, implies that there is a point y_0 in F such that $\|x_0 - y_0\| = \inf_{y \in F} \|x_0 - y\|$.

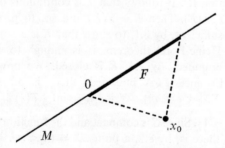

Figure 8.2

8.7 COROLLARY. *Let $f \in C_R[a, b]$, the set of continuous real-valued functions on $[a, b]$, n be a positive integer and M be the set of all real polynomial functions having degree $\leq n$. Then there is a function p_0*

in M such that $\|f-p_0\| = \inf_{p \in M} \|f-p\|$ where $\|\cdot\|$ denotes the uniform norm on $[a, b]$.

Proof. M is a finite-dimensional subspace of $C_R[a, b]$, and so the result follows immediately from 8.6.

8.7 shows that if $f \in C_R[a, b]$ and n is a given positive integer, then, amongst the real polynomial functions of degree $\leqslant n$, there is one which best approximates f in the uniform norm on $[a, b]$. This result is of some significance in numerical analysis.

8.8 THEOREM. *Let f be a continuous function of a connected metric space X into a metric space Y. Them $f(X)$ is a connected subset of Y.*

Proof. Suppose that $f(X)$ is disconnected, and let U, V be open subsets of Y such that $U \cap f(X)$, $V \cap f(X)$ are non-empty, disjoint sets having union $f(X)$. Then, by 7.3, $f^{-1}(U)$ and $f^{-1}(V)$ are open subsets of X. Further, $f^{-1}(U)$ and $f^{-1}(V)$ are non-empty, disjoint sets having union X. This contradicts the hypothesis that X is connected.

8.9 COROLLARY. *Let f be a continuous real-valued function on a connected metric space X, x_1 and x_2 be points of X and k be a real number between $f(x_1)$ and $f(x_2)$. Then there is a point x in X such that $f(x) = k$.*

Proof. $f(X)$ is a connected subset of $\mathbf{R}$, and so, by 4.3, is an interval.

8.2, 8.3 and 8.9 are generalizations of results, usually proved in a first course on analysis, about a continuous real-valued function on a bounded, closed interval $[a, b]$. The proofs of these generalizations which we have given are little more difficult than the traditional proofs of the special cases. Moreover our proofs indicate quite clearly on what properties of the interval $[a, b]$ the special cases depend, namely compactness and connectedness.

§9. Uniform convergence

Let (f_n) be a sequence of complex-valued functions defined on a set X. Suppose that, for each $x \in X$, the sequence $(f_n(x))$ of complex numbers converges. Then we can define a function f on X by setting

$$f(x) = \lim_{n \to \infty} f_n(x) \quad (x \in X).$$

This function f is called the **limit function** of the sequence (f_n), and we say that the sequence (f_n) **converges (pointwise)** on X **to** f.

Suppose that X is a metric space and that each of the functions f_n is continuous at a point x_0 of X. Then we naturally ask whether the limit function f is also continuous at x_0? If x_0 is a cluster point of X, then, using the comment at the end of §7, we see that

f is continuous at x_0

$$\Leftrightarrow \lim_{x \to x_0} f(x) = f(x_0)$$

$$\Leftrightarrow \lim_{x \to x_0} [\lim_{n \to \infty} f_n(x)] = \lim_{n \to \infty} f_n(x_0) = \lim_{n \to \infty} [\lim_{x \to x_0} f_n(x)],$$

each of the functions f_n being continuous at x_0. Expressed in this way, it seems highly plausible that f is continuous at x_0. The following example, however, shows that this need not be the case.

9.1 Example. Let (f_n) be the sequence of functions on $[0, 1]$ defined by

$$f_n(x) = x^n \quad (0 \leqslant x \leqslant 1;\ n = 1, 2, \ldots).$$

Then

$$\lim_{n \to \infty} f_n(x) = \begin{cases} 0 \text{ for } 0 \leqslant x < 1 \\ 1 \text{ for } x = 1. \end{cases}$$

Each of the functions f_n is continuous at 1, but the limit function f of the sequence (f_n) is not continuous at 1.

As the answer to the above question is negative, we are led to study a method of convergence "stronger" than pointwise convergence, namely uniform convergence. We shall show that continuity is preserved under uniform convergence.

As before, let (f_n) be a sequence of complex-valued functions on a set X. The sequence converges pointwise on X to a function f if and only if for each $\varepsilon > 0$ and each $x \in X$, there is a positive integer n_0 such that

$$|f_n(x) - f(x)| < \varepsilon \text{ for all } n \geqslant n_0. \qquad (9.1)$$

When (f_n) converges on X to f, the integers n_0 satisfying (9.1) usually depend on ε and on x. If an integer n_0 can be found, satisfying (9.1) and independent of x, the sequence (f_n) is said to converge uniformly on X to f. More formally:

CONTINUOUS FUNCTIONS

9.2 DEFINITION. The sequence (f_n) is said to **converge uniformly on** X to f if for each $\varepsilon > 0$, there is a positive integer n_0 such that $|f_n(x)-f(x)| < \varepsilon$ for all $x \in X$ and all $n \geq n_0$.

It is easily seen that (f_n) converges uniformly on X to f if and only if $\lim_{n \to \infty} (\sup_{x \in X} |f_n(x)-f(x)|) = 0$.

We note that pointwise convergence on a set does not imply uniform convergence on that set. For example, consider the sequence (f_n) of functions defined in example 9.1. For each positive integer n,

$$\sup_{0 \leq x \leq 1} |f_n(x)-f(x)| \geq |f_n(1-n^{-1})-f(1-n^{-1})| = (1-n^{-1})^n$$

and $\lim (1-n^{-1})^n = e^{-1} \neq 0$. Alternatively, if $\varepsilon > 0$ and $0 < x < 1$, then

$$|f_n(x)-f(x)| < \varepsilon \Leftrightarrow x^n < \varepsilon$$

$$\Leftrightarrow n > \frac{\log \varepsilon}{\log x} \quad (\log x \text{ is negative}):$$

since $\sup \{ \log \varepsilon/\log x : 0 < x < 1 \} = \infty$ (assuming $\varepsilon < 1$), there is no positive integer n_0 such that

$$|f_n(x)-f(x)| < \varepsilon \text{ for all } x \in [0, 1] \text{ and all } n \geq n_0.$$

Our first result gives a criterion for the uniform convergence of a sequence of functions which does not involve any prior knowledge of the limit function of the sequence.

9.3 THEOREM. *Let (f_n) be a sequence of complex-valued functions on a set X. The sequence (f_n) converges uniformly on X if and only if, for each $\varepsilon > 0$, there is a positive integer n_0 such that $|f_p(x)-f_q(x)| < \varepsilon$ for all $x \in X$ and all $p, q \geq n_0$.*

Proof. Suppose that the sequence (f_n) converges uniformly on X to a function f, and let $\varepsilon > 0$. Then there is a positive integer n_0 such that

$$|f_n(x)-f(x)| < \tfrac{1}{2}\varepsilon \text{ for all } x \in X \text{ and all } n \geq n_0.$$

If $p, q \geq n_0$, then, for all $x \in X$,

$$|f_p(x)-f_q(x)| \leq |f_p(x)-f(x)| + |f(x)-f_q(x)| < \tfrac{1}{2}\varepsilon + \tfrac{1}{2}\varepsilon = \varepsilon.$$

Conversely, suppose that the condition of the theorem is satisfied. Then, for each $x \in X$, the sequence $(f_n(x))$ of complex numbers is

Cauchy. Since $\mathbf{C}$ is complete, the sequence $(f_n(x))$ converges for each $x \in X$, and we can define a function f on X by $f(x) = \lim f_n(x)$ ($x \in X$). Let $\varepsilon > 0$. Then, by hypothesis, there is a positive integer n_0 such that $\left| f_p(x) - f_q(x) \right| < \varepsilon$ for all $x \in X$ and all $p, q \geqslant n_0$. Hence, for all $x \in X$ and all $p \geqslant n_0$,

$$\left| f_p(x) - f(x) \right| = \lim_{q \to \infty} \left| f_p(x) - f_q(x) \right| \leqslant \varepsilon.$$

Thus the sequence (f_n) converges uniformly on X to f.

We next show, as promised, that continuity is preserved under uniform convergence.

9.4 THEOREM. *Let (f_n) be a sequence of complex-valued functions on a metric space X which converges uniformly on X to a function f. Suppose that each of the functions f_n is continuous at a point x_0 of X. Then f is continuous at x_0.*

Proof. Let $\varepsilon > 0$. Then there is a positive integer n_0 such that $\left| f_n(x) - f(x) \right| < \frac{1}{3}\varepsilon$ for all $x \in X$ and all $n \geqslant n_0$. Since the function f_{n_0} is continuous at x_0, there is a $\delta > 0$ such that, for all $x \in B(x_0; \delta)$, we have $\left| f_{n_0}(x) - f_{n_0}(x_0) \right| < \frac{1}{3}\varepsilon$. If $x \in B(x_0; \delta)$, then

$$\left| f(x) - f(x_0) \right| \leqslant \left| f(x) - f_{n_0}(x) \right| + \left| f_{n_0}(x) - f_{n_0}(x_0) \right| + \left| f_{n_0}(x_0) - f(x_0) \right|$$
$$< \tfrac{1}{3}\varepsilon + \tfrac{1}{3}\varepsilon + \tfrac{1}{3}\varepsilon = \varepsilon.$$

Thus f is continuous at x_0.

As a corollary to 9.3 and 9.4, we deduce the completeness of the metric space $C[a, b]$, introduced in §1.

9.5 THEOREM. *$C[a, b]$, with the uniform metric, is a complete metric space.*

Proof. Let (f_n) be a Cauchy sequence in $C[a, b]$. Then, for each $\varepsilon > 0$, there is a positive integer n_0 such that

$$\sup_{a \leqslant x \leqslant b} \left| f_p(x) - f_q(x) \right| < \varepsilon \text{ for all } p, q \geqslant n_0.$$

According to 9.3, there is a function f on $[a, b]$ such that (f_n) converges uniformly on $[a, b]$ to f, i.e. such that

$$\lim_{n \to \infty} \sup_{a \leqslant x \leqslant b} \left| f_n(x) - f(x) \right| = 0. \qquad (9.2)$$

CONTINUOUS FUNCTIONS

By 9.4, f is continuous on $[a, b]$ and so belongs to $C[a, b]$. Finally (9.2) asserts that (f_n) converges to f with respect to the uniform metric.

We note that the converse of 9.4 is false in general.

9.6 Example. Let (f_n) be the sequence of functions on $[0, 1]$ defined by

$$f_n(x) = n^2 x^2 (1-x^2)^n \quad (0 \leqslant x \leqslant 1; \; n = 1, 2, \ldots).$$

Then $f_n(0) = 0 = f_n(1)$ and if $0 < x < 1$, we have

$$\frac{f_{n+1}(x)}{f_n(x)} = \frac{(n+1)^2}{n^2}(1-x^2) = (1+n^{-1})^2(1-x^2) \to (1-x^2) < 1.$$

Thus (f_n) converges pointwise on $[0, 1]$ to the zero function. Each of the functions f_n is continuous on $[0, 1]$ and the limit function of the sequence (f_n) is continuous on $[0, 1]$. However (f_n) does not converge uniformly on $[0, 1]$. For

$$\sup_{0 \leqslant x \leqslant 1} |f_n(x)| \geqslant f_n(n^{-1}) = (1-n^{-2})^n$$
$$= (1-n^{-1})^n (1+n^{-1})^n \to 1 \quad (\neq 0).$$

Although the converse of 9.4 is not true in general, the following important partial converse is valid.

9.7 Dini's Theorem. *Let (f_n) be a sequence of continuous real-valued functions on a compact metric space X which converges pointwise on X to a function f. Suppose that f is continuous on X and, for each $x \in X$, the sequence $(f_n(x))$ is decreasing. Then (f_n) converges uniformly on X.*

Proof. For each positive integer n, write

$$g_n(x) = f_n(x) - f(x) \quad (x \in X).$$

Each of the functions g_n is continuous on X and, for each $x \in X$, the sequence $(g_n(x))$ is decreasing and converges to 0. We must show that (g_n) converges uniformly on X to the zero function.

Let $\varepsilon > 0$ and, for each positive integer n, write

$$H_n = \{x \in X : g_n(x) < \varepsilon\}.$$

Then (H_n) is a sequence of open subsets of X, $H_{n+1} \supseteq H_n$

($n=1, 2, \ldots$) and $\bigcup_{\infty}^{n=1} H_n = X$. Since X is compact, there is a positive integer p such that $X = \bigcup_{n=1}^{p} H_n = H_p$. For all $n \geqslant p$ and all $x \in X$, we have $0 \leqslant g_n(x) \leqslant g_p(x) < \varepsilon$. Thus the sequence (g_n) converges uniformly on X to the zero function.

§10. Uniform continuity

Let f be a function of a metric space X into a metric space Y. Then f is continuous on X if and only if it is continuous at each point of X, i.e. if and only if for each $\varepsilon > 0$ and each $x_0 \in X$, there is a $\delta > 0$ such that

$$d_Y(f(x), f(x_0)) < \varepsilon \text{ for all } x \in X \text{ satisfying } d_X(x, x_0) < \delta. \quad (10.1)$$

When f is continuous on X, the numbers δ satisfying (10.1) usually depend on ε and x_0. If a $\delta > 0$ can be found, satisfying (10.1) and independent of x_0, f is said to be uniformly continuous on X. More formally:

10.1 DEFINITION. f is said to be **uniformly continuous on** X if for each $\varepsilon > 0$, there is a $\delta > 0$ such that $d_Y(f(x), f(x_0)) < \varepsilon$ for all $x, x_0 \in X$ satisfying $d_X(x, x_0) < \delta$.

Obviously if f is uniformly continuous on X, then f is continuous on X. As the reader may already know, the converse is false in general.

10.2 *Example.* The function f of $\mathbf{R}$ into $\mathbf{R}$ defined by

$$f(x) = x^2 \quad (x \in \mathbf{R})$$

is certainly continuous on $\mathbf{R}$. However if $x, x_0 \in \mathbf{R}$, $x_0, \delta > 0$ and $|x - x_0| \leqslant \delta$, then

$$|f(x) - f(x_0)| = |x^2 - x_0^2| \leqslant \delta |x + x_0| \leqslant \delta(2x_0 + \delta),$$

and this inequality cannot be improved, since equality actually occurs when $x = x_0 + \delta$. If $\varepsilon > 0$, then

$$\delta(2x_0 + \delta) \leqslant \varepsilon \Leftrightarrow (\delta + x_0)^2 \leqslant \varepsilon + x_0^2$$

$$\Leftrightarrow \delta \leqslant \sqrt{(\varepsilon + x_0^2)} - x_0 = \frac{\varepsilon}{\sqrt{(\varepsilon + x_0^2)} + x_0}.$$

Thus if we want to make $|f(x) - f(x_0)| \leqslant \varepsilon$, the largest value of δ we can take is $\dfrac{\varepsilon}{\sqrt{(\varepsilon+x_0^2)}+x_0}$. Since $\inf\left\{\dfrac{\varepsilon}{\sqrt{(\varepsilon+x_0^2)}+x_0}: x_0 > 0\right\} = 0$,

we cannot find $\delta > 0$ such that

$|f(x)-f(x_0)| \leqslant \varepsilon$ for all $x, x_0 \in \mathbf{R}$ satisfying $|x-x_0| \leqslant \delta$.

Thus f is not uniformly continuous on $\mathbf{R}$.

10.3 THEOREM. *Let f be a continuous function of a compact metric space X into a metric space Y. Then f is uniformly continuous on X.*

Proof. Let $\varepsilon > 0$. Then, for each $x \in X$, there is a $\delta(x) > 0$ such that $d_Y(f(x'), f(x)) < \tfrac{1}{2}\varepsilon$ for all $x' \in X$ satisfying $d_X(x', x) < \delta(x)$. Write

$$B(x) = \{x' \in X : d_X(x', x) < \tfrac{1}{2}\delta(x)\} \quad (x \in X).$$

Then $\{B(x) : x \in X\}$ is an open covering of X, and so, since X is compact, there is a finite set of points $x_1, \ldots, x_n$ in X such that the balls $B(x_1), \ldots, B(x_n)$ cover X. Write $\delta = \tfrac{1}{2} \min \{\delta(x_1), \ldots, \delta(x_n)\}$: $\delta > 0$. Let u, v be any two points of X such that $d_X(u, v) < \delta$. Then there is a positive integer $j, 1 \leqslant j \leqslant n$, such that $u \in B(x_j)$. Since $\delta \leqslant \tfrac{1}{2}\delta(x_j)$,

$$d_X(v, x_j) \leqslant d_X(v, u) + d_X(u, x_j) < \delta + \tfrac{1}{2}\delta(x_j) \leqslant \delta(x_j).$$

Hence

$$d_Y(f(u), f(v)) \leqslant d_Y(f(u), f(x_j)) + d_Y(f(x_j), f(v)) < \tfrac{1}{2}\varepsilon + \tfrac{1}{2}\varepsilon = \varepsilon.$$

The result follows.

As a simple application of 10.3, we deduce a useful result about the approximation of continuous functions by step functions. A real-valued function φ on $\mathbf{R}$ is called a **step function** if there are real numbers $x_0, x_1, \ldots, x_n, v_1, \ldots, v_n$ such that $x_0 < x_1 < \ldots < x_n$, $\varphi(x) = 0$ for $x \in \mathbf{R} \,\, [x_0, x_n]$ and $\varphi(x) = v_i$ for $x_{i-1} < x < x_i$ ($i = 1, \ldots, n$).

10.4 COROLLARY. *Let $f \in C_\mathbf{R}[a, b]$ and $\varepsilon > 0$. Then there is a step function φ such that*

$$\sup_{a \leqslant x \leqslant b} |f(x) - \varphi(x)| < \varepsilon.$$

Proof. By 10.3, there is a $\delta > 0$ such that $|f(x)-f(y)| < \varepsilon$ for all $x, y \in [a, b]$ satisfying $|x-y| < \delta$. Let $x_0, x_1, \ldots, x_n$ be any set of numbers such that $a = x_0 < x_1 < \ldots < x_n = b$ and $\max_{1 \leq i \leq n} (x_i - x_{i-1}) < \delta$, and define φ by

$$\varphi(x) = f(x_{i-1}) \text{ for } x_{i-1} \leq x < x_i \quad (i = 1, \ldots, n),$$
$$\varphi(b) = f(x_{n-1}) \text{ and } \varphi(x) = 0 \text{ for } x \in \mathbf{R} - [a, b].$$

Then φ is a step function and, since

$$|f(x) - \varphi(x)| = |f(x) - f(x_{i-1})| < \varepsilon$$

if $\quad x_{i-1} \leq x < x_i$

and

$$|f(b) - \varphi(b)| = |f(x_n) - f(x_{n-1})| < \varepsilon,$$

we have $\sup_{a \leq x \leq b} |f(x) - \varphi(x)| < \varepsilon$.

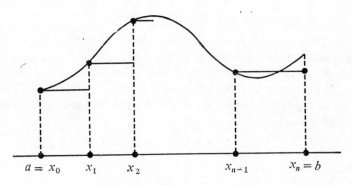

Figure 10.1. The graphs of f and φ.

§11. Weierstrass's Theorem

Let $f \in C_\mathbf{R}[a, b]$ and write

$$d_n = \inf_{p \in P_n} \|f - p\| \quad (n = 1, 2, \ldots)$$

where P_n denotes the set of real polynomial functions of degree $\leq n$ and $\|\cdot\|$ denotes the uniform norm on $[a, b]$. In 8.6, we showed that, for each positive integer n, there is a function p_n in P_n such that $d_n = \|f - p_n\|$. Thus, if f is not a polynomial function, d_n is positive $(n = 1, 2, \ldots)$. In numerical analysis, it is of some interest to obtain estimates for d_n. In this section, we shall be content to prove Weierstrass's theorem which asserts that $\lim d_n = 0$. We require the following preliminary result.

CONTINUOUS FUNCTIONS

11.1 LEMMA. *For each positive integer n and each real number x,*

$$\sum_{k=0}^{n} \binom{n}{k} (nx-k)^2 x^k (1-x)^{n-k} = nx(1-x)$$

where $\binom{n}{k}$ *denotes the binomial coefficient* $\dfrac{n!}{(n-k)!\, k!}$.

Proof. By the binomial theorem,

$$(1+t)^n = \sum_{k=0}^{n} \binom{n}{k} t^k \qquad (t \in \mathbf{R}).$$

Differentiating this identity and multiplying by t, we obtain

$$nt(1+t)^{n-1} = \sum_{k=0}^{k} k \binom{n}{k} t^k.$$

Repetition of this procedure gives

$$nt(1+t)^{n-1} + n(n-1)t^2(1+t)^{n-2} = nt(1+nt)(1+t)^{n-2}$$
$$= \sum_{k=0}^{n} k^2 \binom{n}{k} t^k.$$

Replacing t by $\dfrac{x}{1-x}$ in these three identities and multiplying by $(1-x)^n$, we have

$$\sum_{k=0}^{n} \binom{n}{k} x^k (1-x)^{n-k} = 1 \tag{11.1}$$

$$\sum_{k=0}^{n} k \binom{n}{k} x^k (1-x)^{n-k} = nx \tag{11.2}$$

$$\sum_{k=0}^{n} k^2 \binom{n}{k} x^k (1-x)^{n-k} = nx(1-x+nx). \tag{11.3}$$

Multiplying (11.1) by n^2x^2, (11.2) by $-2nx$ and adding the resulting equations to (11.3), we obtain

$$\sum_{k=0}^{n} (nx-k)^2 \binom{n}{k} x^k (1-x)^{n-k} = nx(nx - 2nx + 1 - x + nx)$$
$$= nx(1-x).$$

11.2 THEOREM. *Let f be a continuous real-valued function on* $[0, 1]$ *and* $\varepsilon > 0$. *Then there is a real polynomial function p such that*

$$\sup_{0 \leq x \leq 1} |f(x) - p(x)| < \varepsilon.$$

48   MATHEMATICAL ANALYSIS

Proof. According to 10.3, f is uniformly continuous on $[0, 1]$. Hence there is a $\delta > 0$ such that $|f(x)-f(y)| < \tfrac{1}{2}\varepsilon$ for all x, y in $[0, 1]$ satisfying $|x-y| < \delta$. For each positive integer n, define B_n by

$$B_n(x) = \sum_{k=0}^{n} f\left(\frac{k}{n}\right)\binom{n}{k} x^k(1-x)^{n-k} \qquad (x \in \mathbf{R}).$$

B_n is a polynomial function of degree at most n: it is called the nth Bernstein polynomial corresponding to the function f. Let $x \in [0, 1]$ and n be a positive integer. Then, using identity (11.1) above,

$$\begin{aligned} |f(x) - B_n(x)| &= \left| \sum_{k=0}^{n} \left[f(x) - f\left(\frac{k}{n}\right)\right]\binom{n}{k} x^k(1-x)^{n-k} \right| \\ &\leq \sum_{k=0}^{n} \left|f(x) - f\left(\frac{k}{n}\right)\right|\binom{n}{k} x^k(1-x)^{n-k} \\ &= (\textstyle\sum' + \sum'') \left|f(x) - f\left(\frac{k}{n}\right)\right|\binom{n}{k} x^k(1-x)^{n-k} \end{aligned}$$

where $\sum'$ denotes the sum over those k for which $\left|x - \dfrac{k}{n}\right| < \delta$ and $\sum''$ denotes the sum over the remaining k. Now

$$\begin{aligned} \sum' \left|f(x) - f\left(\frac{k}{n}\right)\right|\binom{n}{k} x^k(1-x)^{n-k} \\ &< \tfrac{1}{2}\varepsilon \sum' \binom{n}{k} x^k(1-x)^{n-k} \\ &\leq \tfrac{1}{2}\varepsilon \sum_{k=0}^{n} \binom{n}{k} x^k(1-x)^{n-k} = \tfrac{1}{2}\varepsilon, \end{aligned}$$

and, if $M = \sup\limits_{0 \leq x \leq 1} |f(x)|$,

$$\begin{aligned} \sum'' \left|f(x) - f\left(\frac{k}{n}\right)\right|\binom{n}{k} x^k(1-x)^{n-k} \\ &\leq 2M \sum'' \binom{n}{k} \frac{\left(x - \dfrac{k}{n}\right)^2}{\left(x - \dfrac{k}{n}\right)^2} x^k(1-x)^{n-k} \end{aligned}$$

$$\leq \frac{2M}{\delta^2} \sum'' \binom{n}{k} \left(x - \frac{k}{n}\right)^2 x^k (1-x)^{n-k}$$

$$\leq \frac{2M}{\delta^2} \sum_{k=0}^{n} \binom{n}{k} \left(x - \frac{k}{n}\right)^2 x^k (1-x)^{n-k}$$

$$\leq \frac{2M}{\delta^2} \frac{x(1-x)}{n} \qquad \text{by 11.1}$$

$$\leq \frac{M}{2n\delta^2}$$

since $x(1-x) \leq \frac{1}{4}$. Hence, if $n > \dfrac{M}{\varepsilon \delta^2}$, we have

$$|f(x) - B_n(x)| < \tfrac{1}{2}\varepsilon + \frac{M}{2n\delta^2} < \varepsilon$$

for each x in $[0, 1]$.

The above proof is rather difficult, and so some explanation seems called for. First of all, we note that for each $x \in [0, 1]$, $B_n(x)$ is a weighted average of the values of f at the points $0, \dfrac{1}{n}, \dfrac{2}{n}, \ldots, 1$: $f\left(\dfrac{k}{n}\right)$ is given non-negative weight $w_k(x) = \binom{n}{k} x^k (1-x)^{n-k}$ and the sum of the weights is 1. It is easily checked that w_k assumes its largest value at $\dfrac{k}{n}$. Using (11.1), we have

$$|f(x) - B_n(x)| \leq \sum_{k=0}^{n} \left|f(x) - f\left(\frac{k}{n}\right)\right| \binom{n}{k} x^k (1-x)^{n-k}.$$

The terms of the above sum with $\dfrac{k}{n}$ close to x are small since f is continuous at x. The remaining terms are small since the weights $w_k(x)$ are small. As we have to estimate the terms of the sum in different ways, it is natural to split the sum into two parts.

As a simple consequence of 11.2, we have:

11.3 WEIERSTRASS'S THEOREM. *Let f be a continuous real-valued function on an interval $[a, b]$ and $\varepsilon > 0$. Then there is a real polynomial function p such that*

$$\sup_{a \leqslant x \leqslant b} |f(x)-p(x)| < \varepsilon.$$

Proof. Define g on $[0, 1]$ by

$$g(y) = f(a+y(b-a)) \qquad (0 \leqslant y \leqslant 1).$$

Then $g \in C_\mathbf{R}[0, 1]$, and so by 11.2 there is a real polynomial function q such that

$$\sup_{0 \leqslant y \leqslant 1} |g(y)-q(y)| < \varepsilon.$$

If we define p by $p(x) = q\left(\dfrac{x-a}{b-a}\right)$ $(x \in \mathbf{R})$, then p is a real polynomial function and

$$\sup_{a \leqslant x \leqslant b} |f(x)-p(x)| = \sup_{a \leqslant x \leqslant b} \left| g\left(\frac{x-a}{b-a}\right) - q\left(\frac{x-a}{b-a}\right) \right|$$

$$= \sup_{0 \leqslant y \leqslant 1} |g(y)-q(y)| < \varepsilon.$$

§12. The Stone-Weierstrass Theorem

12.1 DEFINITION. A collection A of complex-valued functions on a set E is called a real (complex) **algebra of functions on** E if whenever f, g are functions in A and α is a real (complex) number, the functions $f, g, \alpha f$ and fg also belong to A.

For example, $C_\mathbf{R}[a, b]$ is a real algebra of functions on the interval $[a, b]$. The collection P of all real polynomial functions is also a real algebra of functions on $[a, b]$: it is a subalgebra of $C_\mathbf{R}[a, b]$. Weierstrass's theorem asserts that *the uniform closure of P in $C_\mathbf{R}[a, b]$ is all of $C_\mathbf{R}[a, b]$*.

We may now ask: Just what are the properties of P which make this result true? In this section, we shall obtain an answer to this question. In fact we shall carry out the discussion in a more general setting than is strictly speaking necessary, thereby obtaining a generalization of Weierstrass's result.

Let X be a compact metric space, and denote the set of all continuous complex-valued functions on X by $C(X)$. $C(X)$ is a complex algebra of functions on X. Since X is compact, the functions in $C(X)$ are bounded, and we may define

$$\|f\| = \sup_{x \in X} |f(x)| \qquad (f \in C(X)).$$

CONTINUOUS FUNCTIONS 51

It is easily shown that $\|\cdot\|$ is a norm on $C(X)$, the **uniform norm** on X, and that $C(X)$ is complete with respect to the corresponding metric. The proof of this last assertion is the same as that of 9.5. We denote the subset of $C(X)$ consisting of real-valued functions by $C_\mathbf{R}(X)$: $C_\mathbf{R}(X)$ is a real algebra of functions on X.

12.2 Definition. If f and g are real-valued functions on a set E, we define the functions $f \vee g$ and $f \wedge g$ (read f cup g and f cap g) on E by

$$(f \vee g)(x) = \max\{f(x), g(x)\}, \quad (f \wedge g)(x) = \min\{f(x), g(x)\}.$$

12.3 Definition. A collection L of real-valued functions on a set is called a **lattice** if it is closed under the operations $\vee$ and $\wedge$.

For example, if X is a compact metric space, $C_\mathbf{R}(X)$ is a lattice. The easiest way to see this is to note that if f, g are real-valued functions with the same domain, then

$$f \vee g = \tfrac{1}{2}(f+g+|f-g|) \text{ and } f \wedge g = \tfrac{1}{2}(f+g-|f-g|).$$

12.4 Stone's Theorem. *Let X be a compact metric space and L be a subset of $C_\mathbf{R}(X)$ with the following properties:*
(a) *L is a lattice;*
(b) *if $a, b \in \mathbf{R}$ and x, y are distinct points of X, there is a function f in L such that $f(x) = a, f(y) = b$.*
Then L is uniformly dense in $C_\mathbf{R}(X)$, i.e. if $g \in C_\mathbf{R}(X)$ and $\varepsilon > 0$, there is a function h in L such that $\|g-h\| = \sup_{x \in X} |g(x)-h(x)| < \varepsilon$.

Proof. Let $g \in C_\mathbf{R}(X)$ and $\varepsilon > 0$. Let $x \in X$. Then, for each $y \in X$, there is a function f_{xy} in L such that

$$f_{xy}(x) = g(x) \quad \text{and} \quad f_{xy}(y) = g(y).$$

Since the functions g and f_{xy} are continuous, the set

$$V(y) = \{z \in X : g(z) - f_{xy}(z) < \varepsilon\}$$

is open. Further $y \in V(y)$. Hence $\{V(y) : y \in X\}$ is an open covering of X. (Figure 12.1 illustrates the situation when $X = [a, b]$.) Since X is compact, there is a finite set of points $y_1, y_2, \ldots, y_n$ in X such that the sets $V(y_1), V(y_2), \ldots, V(y_n)$ cover X. Write

$$h_x = f_{xy_1} \vee f_{xy_2} \vee \cdots \vee f_{xy_n}.$$

Then $h_x \in L$ since L is lattice, $h_x(x) = g(x)$ and if $z \in X$, say $z \in V(y_j)$,

$$h_x(z) \geq f_{xy_j}(z) > g(z) - \varepsilon \tag{12.1}$$

by the definition of $V(y_j)$. The set

$$U(x) = \{z \in X : h_x(z) - g(z) < \varepsilon\}$$

is open and contains x. Hence, by the compactness of X, there is a finite set of points $x_1, x_2, \ldots, x_m$ such that the sets $U(x_1), U(x_2), \ldots, U(x_m)$ cover X. Write

$$h = h_{x_1} \wedge h_{x_2} \wedge \ldots \wedge h_{x_m}.$$

Then $h \in L$ and if $z \in X$, say $z \in U(x_i)$,

$$h(z) \leq h_{x_i}(z) < g(z) + \varepsilon$$

by the definition of $U(x_i)$, and, by (12.1),

$$h(z) > g(z) - \varepsilon.$$

Combining these inequalities, we see that

$$|h(z) - g(z)| < \varepsilon \text{ for each } z \in X.$$

The result follows.

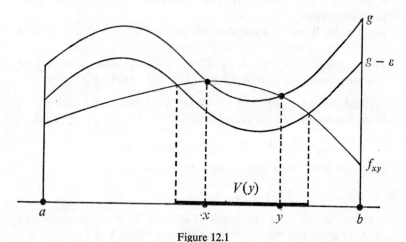

Figure 12.1

Let X be a compact metric space and A be a subalgebra of $C_\mathbf{R}(X)$. Suppose that A is uniformly dense in $C_\mathbf{R}(X)$. Then
  (a) A **separates the points** of X, i.e. if x_1, x_2 are distinct points of X, there is a function f in A such that $f(x_1) \neq f(x_2)$, and

(b) there is no point of X at which all the functions in A vanish. For if there are distinct points x_1, x_2 in X such that $f(x_1) = f(x_2)$ for all $f \in A$, then $g(x_1) = g(x_2)$ for all $g \in C_R(X)$, the uniform closure of A. But $C_R(X)$ contains a function taking distinct values at x_1, x_2: for example, the function h defined by $h(x) = d(x_1, x)$ $(x \in X)$, d being the metric on X, belongs to $C_R(X)$ and $h(x_1) = 0 \neq h(x_2)$. Also if there is a point x_0 in X such that $f(x_0) = 0$ for all $f \in A$, then $g(x_0) = 0$ for all $g \in C_R(X)$: but the real constant functions belong to $C_R(X)$.

We now use Stone's theorem and Weierstrass's theorem to prove that conditions (a) and (b) are sufficient for A to be uniformly dense in $C_R(X)$, thereby obtaining an answer to the question raised about the algebra P of polynomial functions.

12.5 STONE-WEIERSTRASS THEOREM. *Let X be a compact metric space and A be a subalgebra of $C_R(X)$ with the following properties:*
(a) *A separates the points of X;*
(b) *there is no point of X at which all the functions in A vanish.*
Then A is uniformly dense in $C_R(X)$.

Proof. Denote the uniform closure of A in $C_R(X)$ by B. We shall prove that
(1) B is lattice, and
(2) A, and therefore B, satisfies condition (b) of 12.4.
12.4 then implies that $B = C_R(X)$.

(1) Let $f, g \in B$ and $\varepsilon > 0$. We first show that $|f| \in B$. Write $M = \sup_{x \in X} |f(x)|$. According to Weierstrass's theorem, there is a real polynomial function p such that

$$\sup_{-M \leq y \leq M} \big| |y| - p(y) \big| < \tfrac{1}{2}\varepsilon;$$

in particular $|p(0)| < \tfrac{1}{2}\varepsilon$. Write $q = p - p(0)$. Then, for each $y \in [-M, M]$, we have

$$\big| |y| - q(y) \big| \leq \big| |y| - p(y) \big| + |p(y) - q(y)| < \tfrac{1}{2}\varepsilon + \tfrac{1}{2}\varepsilon = \varepsilon.$$

Hence

$$\sup_{x \in X} \big| |f(x)| - q(f(x)) \big| < \varepsilon. \tag{12.2}$$

Now it is easily shown that B is an algebra of functions on X (see Problem 2.9), and so, since $q(0) = 0$, $q \circ f \in B$. (12.2) therefore shows that $|f|$ is in the uniform closure of B in $C_R(X)$. Since B is uniformly closed, it follows that $|f| \in B$. The relations

$$f \vee g = \tfrac{1}{2}(f+g+|f-g|), \quad f \wedge g = \tfrac{1}{2}(f+g-|f-g|)$$

now show that $f \vee g, f \wedge g \in B$. Thus B is a lattice.

(2) Let x, y be distinct points of X and $a, b \in \mathbf{R}$. We first note that there is a function u in A such that $u(x) \neq u(y)$ and $u(x) \neq 0$. For, by hypothesis, there are functions φ and ψ in A such that $\varphi(x) \neq \varphi(y)$ and $\psi(x) \neq 0$. Now $\varphi(x)$ and $\varphi(y)$ cannot both be zero. If $\varphi(x) \neq 0$, we can take $u = \varphi$; if $\varphi(x) = 0$, then $\varphi(y) \neq 0$ and we can take $u = \varphi + \lambda \psi$ where λ is any non-zero real number such that $\lambda \psi(x) \neq \varphi(y) + \lambda \psi(y)$.

Write $\alpha = u(x)(u(x) - u(y))$. Then $\alpha \neq 0$, $g = \alpha^{-1}(u^2 - u(y)u) \in A$, $g(x) = 1$ and $g(y) = 0$. Similarly there is a function h in A such that $h(x) = 0$ and $h(y) = 1$. The function $f = ag + bh$ has the desired properties: $f \in A$, $f(x) = a$ and $f(y) = b$.

In the proof of the above theorem, we did not use the full force of Weierstrass's result. We only used the fact that $|y|$ can be uniformly approximated on an interval $[-M, M]$ by a real polynomial in y. It is worth pointing out that this fact has several fairly simple direct proofs. (See Problems 2.15 and 3.20.)

If X is a compact metric space, it seems natural to ask whether the analogue of 12.5 holds for subalgebras A of $C(X)$, the complex algebra of continuous complex-valued functions on X. Rather surprisingly the answer is negative. (See Problem 2.16.) However, as we shall now prove, the analogue of 12.5 is valid if an additional hypothesis is made about the subalgebra A.

12.6 COMPLEX STONE-WEIERSTRASS THEOREM. *Let X be a compact metric space, and A be a subalgebra of $C(X)$ with the following properties*:
(a) *A separates the points of X;*
(b) *there is no point of X at which all the functions in A vanish;*
(c) *A is **self-adjoint**, i.e. if $f \in A$, then $\bar{f} \in A$.*
Then A is uniformly dense in $C(X)$.

Proof. Denote the set of all real-valued functions in A by $A_\mathbf{R}$: $A_\mathbf{R}$ is a subalgebra of $C_\mathbf{R}(X)$. We first note that $A_\mathbf{R}$ contains the real and imaginary parts of the functions in A. For if $f = u + iv \in A$, u and v being real-valued, then, since A is self-adjoint,

$$u = \tfrac{1}{2}(f+\bar{f}) \in A \quad \text{and} \quad v = \tfrac{1}{2}i\,(\bar{f}-f) \in A.$$

We now show that $A_\mathbf{R}$ satisfies the hypothesis of 12.5.

Let x, y be distinct points of X. Then, since A separates the points of X, there is a function f in A such that $f(x) \ne f(y)$, i.e. $\operatorname{Re} f(x) \ne \operatorname{Re} f(y)$ or $\operatorname{Im} f(x) \ne \operatorname{Im} f(y)$. Since $\operatorname{Re} f$ and $\operatorname{Im} f$ both belong to $A_\mathbf{R}$, it follows that $A_\mathbf{R}$ separates the points of X. By (b), there is a function g in A such that $g(x) \ne 0$, i.e. $\operatorname{Re} g(x) \ne 0$ or $\operatorname{Im} g(x) \ne 0$. Since $\operatorname{Re} g$, $\operatorname{Im} g \in A_\mathbf{R}$, there is no point of X at which all the functions in $A_\mathbf{R}$ vanish. 12.5 now implies that $A_\mathbf{R}$ is uniformly dense in $C_\mathbf{R}(X)$.

Let $f = u+iv \in C(X)$, u, v being real-valued, and $\varepsilon > 0$. As we have shown, there are functions u_1, v_1 in $A_\mathbf{R}$ such that

$$\|u-u_1\| < \tfrac{1}{2}\varepsilon \quad \text{and} \quad \|v-v_1\| < \tfrac{1}{2}\varepsilon.$$

Then $f_1 = u_1 + iv_1 \in A$ and

$$\|f-f_1\| \le \|u-u_1\| + \|i(v-v_1)\| < \tfrac{1}{2}\varepsilon + \tfrac{1}{2}\varepsilon = \varepsilon.$$

The result follows.

§13. Compactness in $(C)X$

At the end of §3, we mentioned that the description of the compact subsets of a metric space is an important, though difficult, problem. In this section, we shall solve this problem for the metric space $C(X)$, X being a compact metric space. We shall require the notion of equicontinuity.

Let (X, d) be a compact metric space. Then any function in $C(X)$ is uniformly continuous on X. Thus if $f \in C(X)$ and $\varepsilon > 0$, there is a $\delta > 0$ such that $|f(x)-f(y)| < \varepsilon$ for all x, y in X satisfying $d(x, y) < \delta$. Let F be a subset of $C(X)$. Then for each $f \in F$ and each $\varepsilon > 0$, there is a $\delta > 0$ such that

$$|f(x)-f(y)| < \varepsilon \text{ for all } x, y \text{ in } X \text{ satisfying } d(x, y) < \delta. \quad (13.1)$$

The numbers δ satisfying (13.1) usually depend on ε and f. If a $\delta > 0$ can be found, satisfying (13.1) and independent of $f \in F$, the family F is said to be equicontinuous on X. More formally:

13.1 DEFINITION. F is said to be **equicontinuous** (on X) if for each $\varepsilon > 0$, there is $\delta > 0$ such that $|f(x)-f(y)| < \varepsilon$ for all f in F and all x, y in X satisfying $d(x, y) < \delta$.

For example, if F is a family of functions on an interval $[a, b]$ and there is a number M such that $|f(x)-f(y)| \leq M|x-y|$ for all f in F and all x, y in $[a, b]$, then F is equicontinuous on $[a, b]$.

13.2 Ascoli-Arzelà Theorem. *A subset F of $C(X)$ is compact if and only if it is bounded, closed and equicontinuous.*

Proof. Suppose first that F is a compact subset of $C(X)$. Then, by 3.8, F is bounded and closed. We have only to show that F is equicontinuous. Let $\varepsilon > 0$. The open balls $\{g \in C(X): \|g-f\| < \tfrac{1}{3}\varepsilon\}$, $f \in F$, cover F. Hence, since F is compact, there is a finite set of functions $f_1, f_2, \ldots, f_n$ in F such that

$$\bigcup_{i=1}^{n} \{g \in C(X): \|g-f_i\| < \tfrac{1}{3}\varepsilon\} \supseteq F. \qquad (13.2)$$

Each of the functions $f_1, f_2, \ldots, f_n$ is uniformly continuous on X. Hence, for each positive integer i, $1 \leq i \leq n$, there is a $\delta_i > 0$ such that

$$|f_i(x)-f_i(y)| < \tfrac{1}{3}\varepsilon \text{ for all } x, y \text{ in } X \text{ satisfying } d(x, y) < \delta_i.$$

Then $\delta = \min\{\delta_1, \delta_2, \ldots, \delta_n\} > 0$. Let $f \in F$ and x, y be any two points of X satisfying $d(x, y) < \delta$. Then, by (13.2), there is a positive integer i, $1 \leq i \leq n$, such that $\|f-f_i\| < \tfrac{1}{3}\varepsilon$. Hence, since $\delta \leq \delta_i$,

$$|f(x)-f(y)| \leq |f(x)-f_i(x)| + |f_i(x)-f_i(y)| + |f_i(y) - f(y)|$$
$$\leq 2\|f-f_i\| + |f_i(x)-f_i(y)|$$
$$< \tfrac{2}{3}\varepsilon + \tfrac{1}{3}\varepsilon = \varepsilon.$$

Thus F is equicontinuous.

Conversely suppose that F is bounded, closed and equicontinuous. To prove that F is compact, it is sufficient, by Problem 1.12, to show that F is sequentially compact, i.e. that every sequence in F has a subsequence which converges to a point of F. Let (g_n) be a sequence in F. Since F is closed, it is enough to show that (g_n) has a subsequence which converges uniformly on X.

We pointed out immediately after Problem 1.12 that a compact metric space has a countable dense subset. Let $x_1, x_2, \ldots$ be an enumeration of such a subset of X. Since F is a bounded subset of $C(X)$, $(g_n(x_1))$ is a bounded sequence of complex numbers, and so has a convergent subsequence. Let (g_{1n}) be a subsequence of (g_n) such that $(g_{1n}(x_1))$ converges. The sequence $(g_{1n}(x_2))$ is bounded,

CONTINUOUS FUNCTIONS

and so has a convergent subsequence. Let (g_{2n}) be a subsequence of (g_{1n}) which converges at x_2. Of course, the sequence $(g_{2n}(x_1))$ is also convergent. Continuing in this way, we obtain, for each positive integer k, a sequence (g_{kn}) such that $(g_{kn}(x_i))$ is convergent for $i = 1, \ldots, k$, and $(g_{(k+1)n})$ is a subsequence of (g_{kn}) $(k = 1, 2, \ldots)$.

Consider the "diagonal" sequence (g_{nn}). Apart from its first $(k-1)$ terms, (g_{nn}) is a subsequence of (g_{kn}), and hence the sequence $(g_{nn}(x_k))$ is convergent. (See. Fig. 13.1.) This is true for $k = 1, 2, \ldots$. We shall prove that (g_{nn}) converges uniformly on X.

$$
\begin{array}{ccccc}
g_{11} & g_{12} & g_{13} & g_{14} & \cdot \\
g_{21} & g_{22} & g_{23} & g_{24} & \cdot \\
g_{31} & g_{32} & g_{33} & g_{34} & \cdot \\
g_{41} & g_{42} & g_{43} & g_{44} & \cdot \\
\cdot & \cdot & \cdot & \cdot & \cdot
\end{array}
$$

Figure 13.1. Each of the functions $g_{33}, g_{44}, \ldots$ occurs in the third row of the above array: further they occur in this order.

Let $\varepsilon > 0$. Since F is equicontinuous, there is a $\delta > 0$ such that $|f(x)-f(y)| < \tfrac{1}{3}\varepsilon$ for all f in F and all x, y in X satisfying $d(x, y) < \delta$. Since $\{x_1, x_2, \ldots\}$ is dense in X, the open balls $\{y \in X: d(x_i, y) < \delta\}$, $i = 1, 2, \ldots$, cover X. Since X is compact, there is a finite set of points $x_1, \ldots, x_\nu$ such that

$$\bigcup_{i=1}^{\nu} \{y \in X: d(x_i, y) < \delta\} = X.$$

As each of the sequences $(g_{nn}(x_i))$, $i = 1, \ldots, \nu$, is convergent, there is a positive integer n_0 such that

$$|g_{pp}(x_i) - g_{qq}(x_i)| < \tfrac{1}{3}\varepsilon \ (i = 1, \ldots, \nu) \text{ for all } p, q \geq n_0.$$

Let $x \in X$. Then there is an integer j, $1 \leq j \leq \nu$, such that $d(x_j, x) < \delta$. Hence, if $p, q \geq n_0$, we have

$$|g_{pp}(x) - g_{qq}(x)| \leq |g_{pp}(x) - g_{pp}(x_j)| + |g_{pp}(x_j) - g_{qq}(x_j)|$$
$$+ |g_{qq}(x_j) - g_{qq}(x)|$$
$$< \tfrac{1}{3}\varepsilon + \tfrac{1}{3}\varepsilon + \tfrac{1}{3}\varepsilon = \varepsilon.$$

It follows by 9.3 that (g_{nn}) converges uniformly on X.

The above proof implies the following corollary, which is of some importance in the theory of differential and integral equations.

C

13.3 COROLLARY. *If F is a bounded, equicontinuous subset of C(X), every sequence in F has a subsequence which converges uniformly on X.*

We shall see an application of this result in the next chapter.

§14. Topological spaces: an aside

In 5.2 and 7.2 we showed that convergence and continuity could be defined in terms of neighbourhoods, and hence in terms of open sets. The concepts of compactness and connectedness were defined in terms of open sets. These comments suggest that much of the material already discussed could have been carried through in a more general setting, namely in a space, which, although it does not have a metric, has a collection of subsets possessing the important properties of the open subsets of a metric space. 2.3 indicates what are the important properties of these open sets. We have motivated the following definition of a topological space.

14.1 DEFINITION. A **topological space** is a pair $(X, \mathcal{T})$ consisting of a non-empty set X and a collection $\mathcal{T}$ of subsets of X with the following properties:
(a) X and $\emptyset$ belong to $\mathcal{T}$;
(b) the intersection of a finite collection of sets from $\mathcal{T}$ is also in $\mathcal{T}$;
(c) the union of any collection of sets from $\mathcal{T}$ is also in $\mathcal{T}$.
$\mathcal{T}$ is called the **topology** of the topological space $(X, \mathcal{T})$, and the sets in $\mathcal{T}$ are called the **open** subsets of this space.

14.2 *Examples.* (a) Every metric space is a topological space in which "open" has the meaning already introduced.

(b) X, any set; $\mathcal{T} = \{X, \emptyset\}$, the **trivial** topology for X.

(c) X, any set; $\mathcal{T}$, the collection of all subsets of X, the **discrete** topology for X.

(d) X, any infinite set; $\mathcal{T}$ consisting of $\emptyset$ and all subsets U of X such that $X - U$ is finite.

(e) $X = \mathbf{R}$; $\mathcal{T}$ consisting of $\mathbf{R}$, $\emptyset$ and all sets of the form (a, ∞), (a, ∞), $a \in \mathbf{R}$.

Many of the definitions and theorems about metric spaces carry over in an obvious way to topological spaces. For example: a subset K of a topological space $(X, \mathcal{T})$ is said to be compact if every open covering of K contains a finite subcovering, i.e. whenever $K \subseteq \bigcup_{i \in I} G_i$, $\{G_i : i \in I\}$ being a collection of sets from $\mathcal{T}$, there is a

CONTINUOUS FUNCTIONS

finite subset J of I such that $K \subseteq \bigcup_{i \in J} G_i$; a function f of a topological space X into a topological space Y is said to be continuous if $f^{-1}(G)$ is an open subset of X for every open subset G of Y, i.e. if $\mathcal{T}_X, \mathcal{T}_Y$ denote the topologies of X, Y, then $f^{-1}(G) \in \mathcal{T}_X$ for every $G \in \mathcal{T}_Y$; if f is a continuous function of a compact topological space X into a topological space Y, then $f(X)$ is a compact subset of Y. The proof of this result is word-for-word the same as that of 8.1.

Problems on Chapter 2

1. Show that a real-valued function f on a metric space X is continuous if and only if the sets $\{x \in X : f(x) > a\}$, $\{x \in X : f(x) < b\}$ are open for all $a, b \in \mathbf{R}$.

2. Let f be a continuous function of a metric space X into a metric space Y. Show that $(f^{-1}(A))^- \subseteq f^{-1}(A^-)$ for each subset A of Y.

3. Let X be a metric space, F be a non-empty closed subset of X and define the function g by

$$g(x) = \inf_{y \in F} d(x, y) \qquad (x \in X).$$

Show that
(a) $|g(x_1) - g(x_2)| \leqslant d(x_1, x_2)$ for all $x_1, x_2 \in X$;
(b) $g(x) = 0$ if and only if $x \in F$.

Deduce that if K is a compact subset of X, disjoint from F, then $\inf \{d(x, y) : x \in K, y \in F\} > 0$. (See Problem 1.9).

4. A real-valued function f on a metric space X is said to be **lower semi-continuous** at a point x_0 of X if for each $\varepsilon > 0$, there is a neighbourhood U of x_0 such that

$$f(x) > f(x_0) - \varepsilon \text{ for all } x \in U.$$

Let f be a lower semi-continuous function on X. Show that
(a) $\{x \in X : f(x) \leqslant a\}$ is a closed subset of X for each $a \in \mathbf{R}$;
(b) if K is a compact subset of X, the set $\{f(x) : x \in K\}$ is bounded below, and there is a point x_0 in K such that

$$f(x_0) = \inf_{x \in K} f(x).$$

5. Let $\{f_i : i \in I\}$ be a family of continuous real-valued functions on a metric space X such that the set $\{f_i(x) : i \in I\}$ is bounded above for each $x \in X$. Show that the function f defined by

$$f(x) = \sup_{i \in I} f_i(x) \quad (x \in X)$$

is lower semi-continuous on X. Must f be continuous on X?

6. Show that the sequence (f_n) of functions of $\mathbf{R}$ into $\mathbf{R}$ defined by

$$f_n(x) = \frac{nx}{1+n^2x^2} \quad (x \in \mathbf{R}; n = 1, 2, \ldots)$$

converges pointwise on $\mathbf{R}$. Is the convergence uniform on $\mathbf{R}$?

7. Show that the sequence (f_n) of functions of $[0, 1]$ into $\mathbf{R}$ defined by

$$f_n(x) = \begin{cases} nx & \text{if } 0 \leqslant x \leqslant 1/n \\ \dfrac{n(1-x)}{n-1} & \text{if } 1/n < x \leqslant 1 \end{cases} \quad (n = 1, 2, \ldots)$$

converges pointwise on $[0, 1]$. Is the convergence uniform on $[0, 1]$?

8. Let (f_n) be the sequence of functions defined by

$$f_n(x) = x e^{-nx} \quad (x \geqslant 0; n = 1, 2, \ldots).$$

Does the sequence (f_n) converge uniformly on $\{x \in \mathbf{R} : x \geqslant 0\}$?

9. Let (f_n), (g_n) be sequences of complex-valued functions on a set X which converge uniformly on X to f, g respectively. Show that
(a) the sequence (f_n+g_n) converges uniformly on X to $f+g$;
(b) if each of the functions f_n, g_n is bounded, the sequence $(f_n g_n)$ converges uniformly on X to fg.
Show that (b) need not be true if the functions f_n, g_n are unbounded.

10. Let $g \in C[a, b]$ and k be a continuous complex-valued function on the square $[a, b] \times [a, b]$ such that

$$(b-a) \sup_{a \leqslant x, y \leqslant b} |k(x, y)| < 1.$$

Show that the function K defined on $C[a, b]$ by

$$(Kf)(x) = g(x) + \int_a^b k(x, y)f(y)\,dy \quad (f \in C[a, b]; a \leqslant x \leqslant b)$$

is a contraction, and deduce that there is a unique function f_0 in $C[a, b]$ such that

$$f_0(x) = g(x) + \int_a^b k(x, y)f_0(y)\,dy \quad (a \leqslant x \leqslant b).$$

11. Let X, Y be metric spaces, with X complete, x_0 an element of X, δ a positive real number and denote the open ball in X having centre x_0 and radius δ by B. Let φ be a function of $B \times Y$ into X, and suppose that
(a) there is a number α in $(0, 1)$ such that

$$d(\varphi(x_1, y), \varphi(x_2, y)) \leqslant \alpha d(x_1, x_2) \text{ for all } x_1, x_2 \text{ in } B \text{ and } y \text{ in } Y$$

and

$$d(\varphi(x_0, y), x_0) < \delta(1-\alpha) \quad \text{for all } y \text{ in } Y,$$

(b) for each x in B, the function $\varphi(x, \cdot)$ is continuous on Y.
Show that, for each y in Y, $\varphi(\cdot, y)$ has a unique fixed point $u(y)$ in B and the function u of Y into B is continuous.
[Hint: use Problem 1.20.]

12. Let f be a real-valued function whose derivative f' is defined and continuous on a neighbourhood of a real number a and satisfies $f'(a) \neq 0$. Show that there is an open neighbourhood U of a such that $V = f(U)$ is an open neighbourhood of $f(a)$, $f \mid U$, the restriction of f to U, is one-one and g, the inverse of $f \mid U$, is continuous on V.
[Hint: consider the function φ defined by $\varphi(x, y) = x - \dfrac{f(x) - y}{f'(a)}$
and use Problem 11.]

13. Let A be a dense subset of a metric space X and f be a function of A into a complete metric space Y. If f is uniformly continuous on A, show that there is a unique continuous function g of X into Y such that $g(a) = f(a)$ for each a in A.

14. If $f \in C_{\mathbf{R}}[0, 1]$ and $\int_0^1 f(x) x^n \, dx = 0$ for $n = 0, 1, 2, \ldots$, show that $f(x) = 0$ for $0 \leqslant x \leqslant 1$.
[Hint: use Weierstrass's theorem.]

15. (a) Let $t \in [0, 1]$ and (x_n) be the sequence defined by

$$x_1 = 0, \ x_{n+1} = x_n + \tfrac{1}{2}(t - x_n^2) \quad (n = 1, 2, \ldots).$$

Show that $0 \leqslant x_n \leqslant \sqrt{t}$ for each positive integer n, and deduce that the sequence (x_n) is increasing and converges to $\sqrt{t}$.
b) Let (p_n) be the sequence of functions defined on $\mathbf{R}$ by

$$p_1(x) = 0, \ p_{n+1}(x) = p_n(x) + \tfrac{1}{2}(x - p_n(x)^2) \ (n = 1, 2, \ldots).$$

Show that p_n is a polynomial function and, using (a), that $(p_n(x))$ converges uniformly on $[0, 1]$ to $\sqrt{x}$.

(c) If $M, \varepsilon > 0$, show that there is a polynomial function p such that
$$\sup_{-M \leq y \leq M} \bigl| |y| - p(y) \bigr| < \varepsilon.$$
[Hint: write $x = y^2/M^2$ and use (b).]

16. Let $X = \{z \in \mathbf{C}: |z| = 1\}$ and A be the subalgebra of $C(X)$ consisting of the restrictions to X of the complex polynomial functions. Show that
(a) A separates the points of X;
(b) there is no point of X at which all the functions in A vanish;
(c) A is not uniformly dense in $C(X)$.

[Hint: in (c), use the fact that $\int_0^{2\pi} f(e^{i\theta}) e^{i\theta}\, d\theta = 0$ for each $f \in A$.]

Why does this example not contradict the Stone-Weierstrass theorem?

17. Let X be a compact metric space and A be a self-adjoint subalgebra of $C(X)$ which separates the points of X. Show that A^-, the uniform closure of A in $C(X)$, either is $C(X)$ or there is a point x_0 in X such that $A^- = \{f \in C(X): f(x_0) = 0\}$.

18. Let k be a continuous complex-valued function on the square $[a, b] \times [a, b]$, and define K on $C[a, b]$ by
$$(Kf)(x) = \int_a^b k(x, y) f(y)\, dy \quad (f \in C[a, b];\ a \leq x \leq b).$$
If $B = \{f \in C[a, b]: \|f\| \leq 1\}$, show that $K(B)^-$ is a compact subset of $C[a, b]$.

19. What are the continuous complex-valued functions on the topological space $(X, \mathscr{T})$ if $\mathscr{T}$ is (a) the trivial topology, (b) the discrete topology?

20. Is it true that every compact subset of a topological space $(X\mathscr{T})$, is closed? If not, what condition on $\mathscr{T}$ will ensure that this result is correct?

CHAPTER 3
FURTHER RESULTS ON UNIFORM CONVERGENCE

§15. Uniform convergence and integration

Let (f_n) be a sequence of bounded real-valued functions on an interval $[a, b]$. Suppose that the sequence (f_n) converges pointwise on $[a, b]$ to a function f and that each of the functions f_n is (Riemann) integrable over $[a, b]$. Two obvious questions arise:
(1) is f integrable over $[a, b]$, and, assuming the answer is yes,
(2) does the sequence $(\int_a^b f_n)$ converge to $\int_a^b f$, i.e. is the equation

$$\lim \int_a^b f_n = \int_a^b (\lim f_n)$$

valid?

The following examples show that the answer to each of these questions can be negative.

15.1 *Examples.* (a) The set of rational numbers in $[0, 1]$ is countable. Let $r_1, r_2, r_3, \ldots$ be an enumeration of this set, and define f_n on $[0, 1]$ by

$$f_n(x) = \begin{cases} 1 & \text{if } x \in \{r_1, r_2, \ldots, r_n\} \\ 0 & \text{otherwise} \end{cases} \quad (n = 1, 2, \ldots).$$

The sequence (f_n) converges pointwise on $[0, 1]$ to the function f given by

$$f(x) = \begin{cases} 1 & \text{if } x \text{ is rational} \\ 0 & \text{otherwise.} \end{cases}$$

Each of the functions f_n is integrable over $[0, 1]$, having only a finite number of discontinuities. However it is easily shown that f is not integrable over $[0, 1]$.

(b) Let (f_n) be the sequence of functions on $[0, 1]$ defined by
$$f_n(x) = nx(1-x^2)^n \quad (0 \leq x \leq 1;\ n = 1, 2, \ldots).$$
Then $f_n(0) = 0 = f_n(1)$ and if $0 < x < 1$, we have
$$\frac{f_{n+1}(x)}{f_n(x)} = \frac{(n+1)}{n}(1-x^2) \to (1-x^2) < 1.$$
Hence $\lim f_n(x) = 0$ for each $x \in [0, 1]$. However
$$\int_0^1 f_n(x)\,dx = \int_0^1 nx(1-x^2)^n\,dx$$
$$= \left[\frac{-n}{2(n+1)}(1-x^2)^{n+1}\right]_0^1 = \frac{n}{2(n+1)} \to \tfrac{1}{2},$$
so that
$$\lim \int_0^1 f_n(x)\,dx \neq \int_0^1 (\lim f_n(x))\,dx.$$

We now show that the answers to each of the questions (1) and (2) is affirmative if the sequence (f_n) converges uniformly on $[a, b]$ to f.

15.2 Theorem. *Let (f_n) be a sequence of bounded real-valued functions on an interval $[a, b]$. Suppose that the sequence (f_n) converges uniformly on $[a, b]$ to a function f and each of the functions f_n is integrable over $[a, b]$. Then f is integrable over $[a, b]$ and*
$$\int_a^b f = \lim \int_a^b f_n. \tag{15.1}$$

Proof. Let $\varepsilon > 0$. Since (f_n) converges uniformly on $[a, b]$ to f, there is a positive integer n_0 such that
$$\sup_{a \leq x \leq b} |f_n(x) - f(x)| \leq \frac{\varepsilon}{3(b-a)} \text{ for all } n \geq n_0.$$
Since f_{n_0} is integrable over $[a, b]$, there is a partition P of $[a, b]$ such that
$$U(f_{n_0}, P) - L(f_{n_0}, P) \leq \tfrac{1}{3}\varepsilon.$$

(If P is the partition $a = x_0 < x_1 < \ldots < x_n = b$ of $[a, b]$, then
$$U(f, P) = \sum_{i=1}^{n} M_i(x_i - x_{i-1}) \quad \text{and} \quad L(f, P) = \sum_{i=1}^{n} m_i(x_i - x_{i-1})$$
where M_i, m_i denote the supremum and infimum of $f([x_{i-1}, x_i])$.)

As $f(x) \leq f_{n_0}(x) + \dfrac{3(b-a)}{\varepsilon}$ for each $x \in [a, b]$, we have
$$U(f, P) \leq U(f_{n_0}, P) + \tfrac{1}{3}\varepsilon.$$

Similarly
$$L(f, P) \geq L(f_{n_0}, P) - \tfrac{1}{3}\varepsilon.$$

Hence
$$\begin{aligned}U(f, P) - L(f, P) &= U(f, P) - U(f_{n_0}, P) + U(f_{n_0}, P) \\ &\quad - L(f_{n_0}, P) + L(f_{n_0}, P) - L(f, P) \leq \varepsilon.\end{aligned}$$

It follows that f is integrable over $[a, b]$.

To prove (15.1), we need only note that for all $n \geq n_0$
$$\left| \int_a^b f - \int_a^b f_n \right| = \left| \int_a^b (f - f_n) \right| \leq \int_a^b |f - f_n| \leq \tfrac{1}{3}\varepsilon.$$

In Chapter 4, when we discuss Lebesgue integration, we shall prove results giving more general conditions under which the order of an integration and a limit operation can be interchanged.

As an application of 15.2 and the Ascoli-Arzelà theorem, we prove that a certain initial value problem has a solution.

15.3 THEOREM. *Let φ be a bounded, continuous real-valued function on the strip $S = \{(x, y) \in \mathbf{R}^2 : 0 \leq x \leq 1\}$ and let c be a real number. Then there is a function f in $C_{\mathbf{R}}[0, 1]$ such that*
$$f(0) = c \text{ and } f'(x) = \varphi(x, f(x)) \quad (0 \leq x \leq 1). \tag{15.2}$$

Proof. We first note that f satisfies (15.2) if and only if
$$f(x) = c + \int_0^x \varphi(t, f(t)) \, dt \qquad (0 \leq x \leq 1). \tag{15.3}$$

This follows easily from the fundamental theorem of calculus. We must show therefore that there is a function f in $C_{\mathbf{R}}[0, 1]$ satisfying the integral equation (15.3).

Let (f_n) be the sequence of functions on $[0, 1]$ defined by

$$f_n(0) = c, \quad f_n(t) = f_n\left(\frac{i}{n}\right) + \varphi\left(\frac{i}{n}, f_n\left(\frac{i}{n}\right)\right)\left(t - \frac{i}{n}\right) \quad \text{for } \frac{i}{n} < t \leq \frac{i+1}{n}$$

$(i = 0, 1, \ldots, n-1)$. (See Figure 15.1.) If $\frac{i}{n} < t < \frac{i+1}{n}$, then $f_n'(t) = \varphi\left(\frac{i}{n}, f_n\left(\frac{i}{n}\right)\right)$, so that, in a sense, f_n is an approximate solution on (15.2). Define Δ_n on $[0, 1]$ by

$$\Delta_n(t) = \begin{cases} f_n'(t) - \varphi(t, f_n(t)) & \text{if } t \text{ is not of the form } \frac{i}{n} \\ 0 & \text{otherwise.} \end{cases}$$

Then, if $x \in [0, 1]$, say $\frac{i}{n} \leq x \leq \frac{i+1}{n}$,

$$f_n(x) = c + \sum_{r=0}^{i-1} \frac{1}{n} \varphi\left(\frac{r}{n}, f_n\left(\frac{r}{n}\right)\right) + \left(x - \frac{i}{n}\right)\varphi\left(\frac{i}{n}, f_n\left(\frac{i}{n}\right)\right)$$

and so we see, splitting the interval $[0, x]$ into the subintervals $\left[0, \frac{1}{n}\right], \left[\frac{1}{n}, \frac{2}{n}\right], \ldots, \left[\frac{i-1}{n}, \frac{i}{n}\right], \left[\frac{i}{n}, x\right]$ and using the fact that $\varphi(t, f_n(t)) + \Delta_n(t)$ has an appropriate constant value on each of the corresponding open intervals, that

$$f_n(x) = c + \int_0^x [\varphi(t, f_n(t)) + \Delta_n(t)] \, dt. \tag{15.4}$$

Our plan is to show that
(a) the sequence (Δ_n) converges uniformly on $[0, 1]$ to the zero function, and
(b) $\{f_1, f_2, \ldots\}$ is a bounded, equicontinuous subset of $C[0, 1]$.
13.3 then implies that the sequence (f_n) has a subsequence (f_{k_n}) which converges uniformly on $[0, 1]$ to a function f. The sequence $(\varphi(t, f_{k_n}(t)))$ converges to $\varphi(t, f(t))$ uniformly on $[0, 1]$. (Prove this.) Hence, replacing f_n in (15.4) by f_{k_n}, we see, using (15.2), that f satisfies (15.3). Now we have only to fill in the details.

To prove (b), write $M = \sup_{(x, y) \in S} |\varphi(x, y)|$. It is then clear from Figure 15.1 that, for all $s, t \in [0, 1]$,

$$|f_n(t)| \leq M + |c| \quad \text{and} \quad |f_n(s) - f_n(t)| \leq M|s-t|$$

$(n = 1, 2, \ldots)$. (b) follows.

To prove (a), let $\varepsilon > 0$. Since φ is uniformly continuous on the rectangle $R = \{(x, y) \in \mathbf{R}^2 : 0 \leq x \leq 1, |y| \leq M + |c|\}$, there is a $\delta > 0$ such that $|\varphi(x_1, y_1) - \varphi(x_2, y_2)| < \varepsilon$ for all (x_1, y_1), (x_2, y_2) in R satisfying $(x_1 - x_2)^2 + (y_1 - y_2)^2 < \delta^2$. Suppose that $t \in \left(\dfrac{i}{n}, \dfrac{i+1}{n}\right)$.

Then

$$\left(\frac{i}{n} - t\right)^2 + \left(f_n\left(\frac{i}{n}\right) - f_n(t)\right)^2 \leq (1 + M^2)\left(\frac{i}{n} - t\right)^2 < \frac{1 + M^2}{n^2}.$$

Hence, if $\dfrac{1 + M^2}{n^2} < \delta^2$, i.e. if $n > \dfrac{\sqrt{(1 + M^2)}}{\delta}$, we have

$$|\Delta_n(t)| = |f_n'(t) - \varphi(t, f_n(t))| = \left|\varphi\left(\frac{i}{n}, f_n\left(\frac{i}{n}\right)\right) - \varphi(t, f_n(t))\right| < \varepsilon.$$

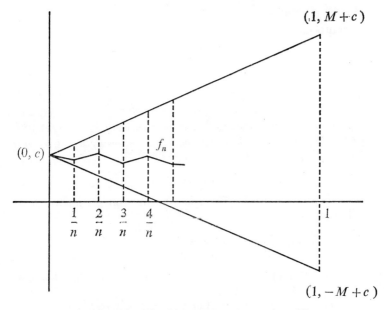

Figure 15.1. The graph of f_n which consists of a number of line segments each having slope between $\pm M = \pm \sup_{(x, y) \in S} |\varphi(x, y)|$.

Since this inequality is automatically satisfied at points of the form $\frac{i}{n}$, (a) follows.

§16. Uniform convergence and differentiation

Theorems 9.4 and 15.2 show that continuity and integrability are preserved under uniform convergence. It is tempting to conjecture that differentiability is also preserved under uniform convergence, that is if (f_n) is a sequence of real-valued functions on a neighbourhood U of a real number x_0, which converges uniformly on U to a function f, and each of the functions f_n is differentiable at x_0, then f is differentiable at x_0 and $f'(x_0) = \lim_{n \to \infty} f_n'(x_0)$, i.e.

$$\lim_{h \to 0}\left(\lim_{n \to \infty} \frac{f_n(x_0+h)-f_n(x_0)}{h}\right) = \lim_{n \to \infty}\left(\lim_{h \to 0} \frac{f_n(x_0+h)-f_n(x_0)}{h}\right).$$

However this result cannot be true. For, by Weierstrass's theorem, the function f given by

$$f(x) = |x| \quad (x \in \mathbf{R})$$

is the uniform limit on $[-1, 1]$, say, of a sequence of polynomial functions, but f is not differentiable at 0. The relation between uniform convergence and differentiation is rather more involved.

16.1 THEOREM. *Let (f_n) be a sequence of real-valued differentiable functions on an open interval (a, b). Suppose that (f_n') converges uniformly on (a, b) to a function g and there is a point x_0 in (a, b) such that $(f_n(x_0))$ converges. Then (f_n) converges uniformly on (a, b) to a function f, f is differentiable on (a, b) and $f' = g$.*

Proof. Let $\varepsilon > 0$. We use 9.3 to show that the sequence (f_n) converges uniformly on (a, b). There is a positive integer n_0 such that for all $p, q \geqslant n_0$

$$|f_p(x_0)-f_q(x_0)| < \tfrac{1}{2}\varepsilon \text{ and } \sup_{a<t<b}|f_p'(t)-f_q'(t)| < \frac{\varepsilon}{2(b-a)}.$$

Let $x, y \in (a, b)$ and $p, q \geqslant n_0$. Then, by the mean value theorem,

$$f_p(x)-f_q(x)-f_p(y)+f_q(y) = (x-y)[f_p'(t)-f_q'(t)]$$

for some number t between x and y, so that

$$|f_p(x)-f_q(x)-f_p(y)+f_q(y)| \leq |x-y|\frac{\varepsilon}{2(b-a)} < \tfrac{1}{2}\varepsilon. \quad (16.1)$$

Hence
$$|f_p(x)-f_q(x)| \leq |f_p(x)-f_q(x)-f_p(x_0)+f_q(x_0)| + |f_p(x_0)-f_q(x_0)|$$
$$< \tfrac{1}{2}\varepsilon+\tfrac{1}{2}\varepsilon = \varepsilon.$$

It follows that the sequence (f_n) converges uniformly on (a, b) to a function f. We have to show that f is differentiable on (a, b) and $f' = g$.

Let $c \in (a, b)$ and define φ_n, $n = 1, 2, \ldots$, and ψ on (a, b) by

$$\varphi_n(x) = \frac{f_n(x)-f_n(c)}{x-c} \quad (x \neq c), \quad \varphi_n(c) = f_n'(c)$$

and
$$\psi(x) = \lim_{n \to \infty} \varphi_n(x).$$

Since $\lim_{x \to c} \varphi_n(x) = f_n'(c) = \varphi_n(c)$, we see that φ_n is continuous at c.

If $p, q \geq n_0$, then, for each $x \in (a, b) - \{c\}$,

$$|\varphi_p(x)-\varphi_q(x)| = \frac{1}{|x-c|}|f_p(x)-f_p(c)-f_q(x)+f_q(c)|$$

$$\leq \frac{\varepsilon}{2(b-a)} \quad \text{by (16.1)}$$

and
$$|\varphi_p(c)-\varphi_q(c)| = |f_p'(c)-f_q'(c)| < \tfrac{1}{2}\varepsilon.$$

Thus the sequence (φ_n) converges uniformly on (a, b). As uniform convergence preserves continuity, ψ is continuous at c, i.e.

$$\lim_{x \to c} \psi(x) = \psi(c) = \lim_{n \to \infty} \varphi_n(c) = \lim_{n \to \infty} f_n'(c) = g(c).$$

But, for $x \neq c$,

$$\psi(x) = \lim_{n \to \infty} \frac{f_n(x)-f_n(c)}{x-c} = \frac{f(x)-f(c)}{x-c}.$$

Hence $\lim_{x \to c} \frac{f(x)-f(c)}{x-c}$ exists and equals $g(c)$, i.e. f is differentiable at c and $f'(c) = g(c)$. This is true for each $c \in (a, b)$.

We remark that if each of the functions f_n' is assumed to be continuous on (a, b), it is possible to give a much simpler proof of the above theorem using 15.2 and the fundamental theorem of calculus. (See Problem 3.3.)

§17. Uniform convergence of series

So far we have only discussed the uniform convergence of sequences of functions on a set. The corresponding definition for series of functions is given in the obvious way.

17.1 DEFINITION. Let (f_n) be a sequence of complex-valued functions on a set X, and, for each positive integer n, write

$$S_n(x) = f_1(x) + f_2(x) + \ldots + f_n(x) \qquad (x \in X). \qquad (17.1)$$

Then the series $\sum_{n=1}^{\infty} f_n$ is said to converge uniformly on X if the sequence (S_n) converges uniformly on X.

9.3 implies the following criterion for the uniform convergence of a series.

17.2 THEOREM. *Let (f_n) be a sequence of complex-valued functions on a set X. Then the series $\sum_{n=1}^{\infty} f_n$ converges uniformly on X if and only if for each $\varepsilon > 0$ there is a positive integer n_0 such that*

$$\left| \sum_{n=p+1}^{q} f_n(x) \right| < \varepsilon \text{ for all } x \text{ in } X \text{ and all } q > p \geqslant n_0.$$

Proof. Define the sequence (S_n) of functions on X by (17.1). According to 9.3, the sequence (S_n) converges uniformly on X if and only if for each $\varepsilon > 0$, there is a positive integer n_0 such that

$$|S_p(x) - S_q(x)| < \varepsilon \text{ for all } x \text{ in } X \text{ and all } p, q \geqslant n_0.$$

Since $|S_p(x) - S_q(x)| = \left| \sum_{n=p+1}^{q} f_n(x) \right|$ if $q > p$, the result follows.

The next result gives the analogue of 9.4 for series of functions.

17.3 THEOREM. *Let (f_n) be a sequence of complex-valued functions on a metric space X. Suppose that the series $\sum_{n=1}^{\infty} f_n$ converges uni-*

formly on X to S and each of the functions f_n is continuous at a point x_0 of X. Then S is continuous at x_0.

Proof. Define (S_n) by (17.1). Since each of the functions f_n is continuous at x_0, S_n is continuous at x_0 $(n = 1, 2, \ldots)$. The result therefore follows from 9.4.

In a similar way, we can prove the following analogues of 15.2 and 16.1.

17.4 THEOREM. *Let (f_n) be a sequence of bounded real-valued functions on an interval $[a, b]$. Suppose that the series $\sum_{n=1}^{\infty} f_n$ converges uniformly on $[a, b]$ to S and each of the functions f_n is integrable over $[a, b]$. Then S is integrable over $[a, b]$ and*

$$\int_a^b S = \sum_{n=1}^{\infty} \int_a^b f_n.$$

17.5 THEOREM. *Let (f_n) be a sequence of real-valued differentiable functions on an open interval (a, b). Suppose that the series $\sum_{n=1}^{\infty} f_n'$ converges uniformly on (a, b) to a function G and there is a point x_0 in (a, b) such that the series $\sum_{n=1}^{\infty} f_n(x_0)$ converges. Then the series $\sum_{n=1}^{\infty} f_n$ converges uniformly on (a, b) to a function F, F is differentiable on (a, b) and $F' = G$.*

17.4, 17.5 give conditions under which it is permissible to integrate and differentiate a series of functions term-by-term.

We note that these results are also valid for sequences of complex-valued functions, provided integrals and derivatives are interpreted in terms of real and imaginary parts.

§18. Tests for uniform convergence of series

If (f_n) is a sequence of complex-valued functions on a set X, it is not normally possible to establish the uniform convergence of the series $\sum_{n=1}^{\infty} f_n$ on X directly from Definition 17.1: it is usually impossible to express

$$S_n(x) = f_1(x) + f_2(x) + \ldots + f_n(x)$$

in closed form. Further the criterion contained in 17.2 is often difficult to apply. Therefore we shall prove some useful tests for uniform convergence of series.

18.1 WEIERSTRASS'S TEST. *Let (f_n) be a sequence of complex-valued functions on a set X, and suppose that there is a convergent series $\sum_{n=1}^{\infty} M_n$ of positive real numbers such that $|f_n(x)| \leqslant M_n$ for each x in X and each positive integer n. Then the series $\sum_{n=1}^{\infty} f_n$ converges uniformly on X.*

Proof. Let $\varepsilon > 0$. Since the series $\sum_{n=1}^{\infty} M_n$ is convergent, there is a positive integer n_0 such that $\sum_{n=p+1}^{q} M_n < \varepsilon$ for all $q > p \geqslant n_0$. Hence, if $q > p \geqslant n_0$, we have for each x in X

$$\left| \sum_{n=p+1}^{q} f_n(x) \right| \leqslant \sum_{n=p+1}^{q} |f_n(x)| \leqslant \sum_{n=p+1}^{q} M_n < \varepsilon.$$

The result therefore follows from 17.2.

The other two tests, which we shall discuss, can be deduced from the following simple lemma.

18.2 LEMMA. *Let (a_n) be a decreasing sequence of non-negative real numbers and (b_n) be any sequence of complex numbers. Then, if p, q are positive integers with $q > p$,*

$$\left| \sum_{n=p+1}^{q} a_n b_n \right| \leqslant a_{p+1} \max_{p+1 \leqslant s \leqslant q} \left| \sum_{n=p+1}^{s} b_n \right|.$$

Proof. Write

$$B_s = \sum_{n=p+1}^{s} b_n \quad \text{and} \quad M = \max\{|B_{p+1}|, |B_{p+2}|, \ldots, |B_q|\}.$$

Then

$$\sum_{n=p+1}^{q} a_n b_n = a_{p+1} B_{p+1} + \sum_{n=p+2}^{q} a_n (B_n - B_{n-1})$$

$$= a_q B_q + \sum_{n=p+1}^{q-1} B_n (a_n - a_{n+1})$$

rearranging terms, so that

UNIFORM CONVERGENCE 73

$$\left| \sum_{n=p+1}^{q} a_n b_n \right| \leq \left| a_q B_q \right| + \sum_{n=p+1}^{q-1} \left| B_n(a_n - a_{n+1}) \right|$$

$$\leq M a_q + \sum_{n=p+1}^{q-1} M(a_n - a_{n+1})$$

$$= M a_{p+1}.$$

18.3 ABEL'S TEST. *Let (f_n) be a sequence of complex-valued functions on a set X and (g_n) be a decreasing sequence of non-negative functions on X. Suppose that*

(a) *the series $\sum_{n=1}^{\infty} f_n$ converges uniformly on X;*
(b) *the sequence (g_n) is **uniformly bounded** on X, i.e. there is a number M such that $|g_n(x)| \leq M$ for each x in X and each positive integer n.*

Then the series $\sum_{n=1}^{\infty} f_n g_n$ converges uniformly on X.

Proof. Let $\varepsilon > 0$. Then, by (a), there is a positive integer n_0 such that

$$\left| \sum_{n=q+1}^{q} f_n(x) \right| < \frac{\varepsilon}{M} \quad \text{for all } x \text{ in } X \text{ and all } q > p \geq n_0.$$

Hence, if $q > p \geq n_0$, we have, using the above lemma, for each x in X

$$\left| \sum_{n=p+1}^{q} f_n(x) g_n(x) \right| \leq g_{p+1}(x) \max_{p+1 \leq s \leq q} \left| \sum_{n=p+1}^{s} f_n(x) \right|$$

$$< M \cdot \frac{\varepsilon}{M} = \varepsilon.$$

The result now follows by 17.2.

18.4 DIRICHLET'S TEST. *Let (f_n) be a sequence of complex-valued functions on a set X and (g_n) be a decreasing sequence of non-negative functions on X. Suppose that*

(a) *the sequence of partial sums of the series $\sum_{n=1}^{\infty} f_n$ is uniformly bounded on X;*
(b) *the sequence (g_n) converges uniformly on X to the zero function.*

Then the series $\sum_{n=1}^{\infty} f_n g_n$ converges uniformly on X.

Proof. According to (a), there is a number M such that

$$\left| \sum_{n=1}^{p} f_n(x) \right| \leq M \quad \text{for each } x \text{ in } X \text{ and each positive integer } p.$$

Let $\varepsilon > 0$. By (b), there is a positive integer n_0 such that

$$|g_n(x)| < \frac{\varepsilon}{2M} \text{ for all } x \text{ in } X \text{ and all } n \geq n_0.$$

Hence, if $q > p \geq n_0$, we have for each x in X

$$\left| \sum_{n=p+1}^{q} f_n(x) g_n(x) \right|$$

$$\leq g_{p+1}(x) \max_{p+1 \leq s \leq q} \left| \sum_{n=p+1}^{s} f_n(x) \right|$$

$$\leq g_{p+1}(x) \max_{p+1 \leq s \leq q} \left\{ \left| \sum_{n=1}^{s} f_n(x) \right| + \left| \sum_{n=1}^{p} f_n(x) \right| \right\}$$

$$< \frac{\varepsilon}{2M}(M+M) = \varepsilon.$$

The result follows.

18.5 *Examples.* (a) The series $\sum_{n=1}^{\infty} n^{-2} x^n$ converges uniformly on $[-1, 1]$.

For each x in $[-1, 1]$, $|n^{-2} x^n| \leq n^{-2}$ ($n = 1, 2, \ldots$) and the series $\sum_{n=1}^{\infty} n^{-2}$ is convergent. Weierstrass's test therefore implies the conclusion.

(b) The series $\sum_{n=1}^{\infty} \frac{(-1)^{n+1}}{n} |x|^n$ converges uniformly on $[-1, 1]$.

For each x in $[-1, 1]$, $(|x|^n)$ is a decreasing sequence of non-negative numbers and $|x|^n \leq 1$ for each positive integer n. Further the series $\sum_{n=1}^{\infty} \frac{(-1)^{n+1}}{n}$ is convergent. Hence Abel's test gives the conclusion.

We note that, since $\sup_{-1 \leq x \leq 1} \left| \frac{(-1)^{n+1}}{n} |x|^n \right| = \frac{1}{n}$ and the series $\sum_{n=1}^{\infty} n^{-1}$ is divergent, Weierstrass's test is not applicable in this example.

(c) Let $\delta \in (0, \pi)$. Then the series $\sum_{n=1}^{\infty} \dfrac{\sin nx}{n}$ converges uniformly on $[\delta, 2\pi - \delta]$.

For each positive integer p and each x in $[\delta, 2\pi - \delta]$,

$$\sum_{n=1}^{p} \sin nx = \frac{1}{2 \sin \tfrac{1}{2}x} \sum_{n=1}^{p} [\cos(n - \tfrac{1}{2})x - \cos(n + \tfrac{1}{2})x]$$

$$= \frac{\cos \tfrac{1}{2}x - \cos(p + \tfrac{1}{2})x}{2 \sin \tfrac{1}{2}x},$$

so that

$$\left| \sum_{n=1}^{p} \sin nx \right| \leq \frac{1}{\sin \tfrac{1}{2}x} \leq \frac{1}{\sin \tfrac{1}{2}\delta}.$$

Further the sequence $(1/n)$ is decreasing and converges to 0. Hence we may apply Dirichlet's test.

We note that although the series $\sum_{n=1}^{\infty} \dfrac{\sin nx}{n}$ converges for each $x \in [0, 2\pi]$, it does *not* converge uniformly on this interval. (See Problem 3.7.)

§19. Power series

For each non-negative integer n, let p_n be the function on **C** given by $p_n(z) = z^n$. Then a series of functions of the form $\sum_{n=1}^{\infty} a_n p_n$, where $a_0, a_1, a_2, \ldots$ is a given sequence of complex numbers, is called a **power series**.

Associated with each power series, there is a circle in the complex plane with centre O, called the **circle of convergence**, such that the series is absolutely convergent at each point inside the circle and diverges at each point outside the circle. The radius of this circle is called the **radius of convergence** of the series. (The circle of convergence may consist of the single point O or may be the whole complex plane. In these cases, it is conventional to say that the radius of convergence is 0 and $+\infty$ respectively.)

In our first theorem, we establish the existence of the circle of convergence and describe a way of calculating its radius. Before stating the result, we recall the definition of the upper limit (lim sup) of a sequence of real numbers. Let (a_n) be a sequence of real numbers. If the sequence is bounded above, write $b_n = \sup_{i \geq n} a_i$ ($n = 1, 2, \ldots$). The sequence (b_n) is decreasing, and the upper limit of (a_n) is defined by $\limsup a_n = \lim b_n$. If the sequence (a_n) is not bounded above, we define $\limsup a_n = +\infty$. (The lower limit (lim inf) of the sequence (a_n) is defined in a similar way.) If $\lambda = \limsup a_n$ is a real number, then it is characterized by the following two conditions:

(1) for each $\varepsilon > 0$, there is a positive integer n_0 such that

$$a_n < \lambda + \varepsilon \text{ for all } n \geq n_0,$$

(2) for each $\varepsilon > 0$ and each positive integer p, there is a positive integer q such that

$$q > p \text{ and } a_q > \lambda - \varepsilon.$$

19.1 THEOREM. *Let $\sum_{n=0}^{\infty} a_n p_n$ be a given power series, and write*

$$R = 1/\rho \text{ where } \rho = \limsup \sqrt[n]{|a_n|}$$

(with the convention that $R = 0$ if $\rho = +\infty$ and $R = +\infty$ if $\rho = 0$). Then the series converges absolutely at each point inside the circle $\{z \in \mathbf{C}: |z| = R\}$ and diverges at each point outside this circle. Further, if $r < R$, the series is uniformly convergent on the disc $\{z \in \mathbf{C}: |z| \leq r\}$.

Proof. Suppose first that $|z| < R = 1/\rho$, and let t be a number satisfying $|z| < t < 1/\rho$. Then $1/t > \rho = \limsup \sqrt[n]{|a_n|}$, and so there is a positive integer n_0 such that $\sqrt[n]{|a_n|} < 1/t$ for all $n \geq n_0$. For all $n \geq n_0$, $|a_n z^n| \leq \left(\dfrac{|z|}{t}\right)^n$. It follows, by the comparison test, that the series $\sum_{n=0}^{\infty} a_n z^n$ is absolutely convergent.

Suppose now that $|z| > R = 1/\rho$. Then $1/|z| < \rho$ and there are infinitely many positive integers n such that $\sqrt[n]{|a_n|} > 1/|z|$ i.e. such that $|a_n z^n| > 1$. Thus the sequence $(a_n z^n)$ does not converge to 0, which implies that the series $\sum_{n=0}^{\infty} a_n z^n$ is divergent.

To prove the final assertion, we note that if $|z| \leq r < R$, then

$$|a_n z^n| \leq |a_n| r^n \quad (n = 0, 1, 2, \ldots).$$

Since the series $\sum_{n=0}^{\infty} a_n r^n$ is absolutely convergent, Weierstrass's test implies that the series $\sum_{n=0}^{\infty} a_n p_n$ converges uniformly on the disc $\{z \in \mathbf{C} : |z| \leq r\}$.

Let $\sum_{n=0}^{\infty} a_n p_n$ be a power series with non-zero radius of convergence R, and define

$$f(z) = \sum_{n=0}^{\infty} a_n z^n$$

for all z for which the series converges. The domain of f contains the open disc $D = \{z \in \mathbf{C} : |z| < R\}$ and is contained in the closure D^- of D. Using the results already obtained about uniformly convergent series, we can deduce several properties of f.

19.2 THEOREM. *f is continuous on D.*

Proof. Let $z_0 \in D$, and choose r such that $|z_0| < r < R$. Then the series $\sum_{n=0}^{\infty} a_n p_n$ converges uniformly on $\{z \in \mathbf{C} : |z| \leq r\}$, and the continuity of f at z_0 therefore follows from 17.3.

If a point z_0 on the boundary of D is in the domain of f, it is natural to ask whether f is continuous at z_0. The answer is negative in general, though we shall not attempt to produce an example. We shall, however, show that f is continuous at z_0 in a rather weak sense, namely that $f(z)$ tends to $f(z_0)$ as z tends *radially* to z_0. By a rotation of axes, it is sufficient to prove this result in the case in which $z_0 = R$.

19.3 ABEL'S THEOREM. *Suppose R is in the domain of f. Then*

$$\lim_{x \to R-} f(x) = f(R).$$

Proof. For each $x \in [0, R]$, $((x/R)^n)$ is a decreasing sequence of non-negative numbers and $(x/R)^n \leq 1$ for each positive integer n.

Further, by hypothesis, the series $\sum_{n=0}^{\infty} a_n R^n$ is convergent. Hence, by 18.3, the series $\sum_{n=0}^{\infty} a_n R^n (x/R)^n = \sum_{n=0}^{\infty} a_n x^n$ converges uniformly on $[0, R]$. Thus f is continuous on $[0, R]$, and the result follows.

19.4 COROLLARY. *Let* $\sum_{n=0}^{\infty} a_n$, $\sum_{n=0}^{\infty} b_n$ *be convergent series, and write*

$$c_n = \sum_{k=0}^{n} a_k b_{n-k} \quad (n = 0, 1, 2, \ldots).$$

Suppose the series $\sum_{n=0}^{\infty} c_n$, *called the Cauchy product of the given two series, is convergent. Then*

$$\sum_{n=0}^{\infty} c_n = \left(\sum_{n=0}^{\infty} a_n \right) \left(\sum_{n=0}^{\infty} b_n \right).$$

Proof. By hypothesis, each of the power series $\sum_{n=0}^{\infty} a_n p_n$, $\sum_{n=0}^{\infty} b_n p_n$, $\sum_{n=0}^{\infty} c_n p_n$ has radius of convergence at least 1. Further the third of these series is the Cauchy product of the first two. Hence, using a result about multiplication of absolutely convergent series,†

$$\sum_{n=0}^{\infty} c_n x^n = \left(\sum_{n=0}^{\infty} a_n x^n \right) \left(\sum_{n=0}^{\infty} b_n x^n \right) \text{ for each } x \in (-1, 1). \quad (19.1)$$

Letting $x \to 1-$ in (19.1) and using 19.3, we obtain the stated result.

19.5 THEOREM. *Let x be a real number in D. Then*

$$\int_0^x f = \sum_{n=0}^{\infty} \frac{a_n x^{n+1}}{(n+1)}.$$

Proof. If r is chosen so that $|x| < r < R$, then the series $\sum_{n=0}^{\infty} a_n p_n$ converges uniformly on $[-r, r]$ to f. Hence, by 17.4,

$$\int_0^x f = \sum_{n=0}^{\infty} \int_0^x a_n p_n = \sum_{n=0}^{\infty} \frac{a_n x^{n+1}}{(n+1)}.$$

† See, for example, J. A. Anderson, *Real Analysis*, Logos Press, London: p. 161.

As an elementary application of 19.3 and 19.5, we have:

19.6 Example. The power series $\sum_{n=0}^{\infty} (-1)^n p_n$ has radius of convergence 1, and

$$\sum_{n=0}^{\infty} (-1)^n z^n = \frac{1}{1+z} \quad \text{for} \quad |z| < 1.$$

Hence, by 19.5,

$$\sum_{n=0}^{\infty} \frac{(-1)^n x^{n+1}}{(n+1)} = \int_0^x \frac{dt}{1+t} = \log(1+x) \text{ for each } x \in (-1, 1).$$

To see that this result is also valid when $x = 1$, we note that the series $\sum_{n=0}^{\infty} \frac{(-1)^n}{(n+1)}$ is convergent, and so, by Abel's theorem,

$$\sum_{n=0}^{\infty} \frac{(-1)^n}{(n+1)} = \lim_{x \to 1-} \log(1+x) = \log 2.$$

19.7 Theorem. *The power series* $\sum_{n=1}^{\infty} n a_n p_{n-1}$ *has radius of convergence R. Further, if x is a real number in D, then f is differentiable at x and*

$$f'(x) = \sum_{n=1}^{\infty} n a_n x^{n-1}.$$

Proof. The power series $\sum_{n=1}^{\infty} n a_n p_{n-1}$ has radius of convergence $1/\sigma$ where $\sigma = \limsup \sqrt[n]{(n|a_n|)}$. Let $\varepsilon > 0$, and write $\eta = \min\{1, \varepsilon/(\rho+2)\}$ where $\rho = \limsup \sqrt[n]{|a_n|} = 1/R$. Since $\lim \sqrt[n]{n} = 1$, there is a positive integer n_0 such that

$$\sqrt[n]{n} < 1 + \eta \quad \text{for all } n \geq n_0.$$

We may also assume that

$$\sqrt[n]{|a_n|} < \rho + \eta \quad \text{for all } n \geq n_0.$$

Then, if $n \geq n_0$,

$$\sqrt[n]{(n|a_n|)} < (\rho+\eta)(1+\eta)$$
$$= \rho + (\rho+1)\eta + \eta^2 \leq \rho + (\rho+2)\eta \leq \rho + \varepsilon.$$

Further there are infinitely many positive integers n such that

$$\sqrt[n]{|a_n|} > \rho - \varepsilon.$$

Since $\sqrt[n]{n} \geq 1$ for each n, there are infinitely many n such that

$$\sqrt[n]{(n|a_n|)} > \rho - \varepsilon.$$

Hence, by the characterization of an upper limit mentioned above, $\sigma = \rho = 1/R$.

The differentiation result follows from 17.5.

19.8 COROLLARY. *f is infinitely differentiable at each point x of $D \cap \mathbf{R}$ and, for each positive integer k,*

$$f^{(k)}(x) = \sum_{n=k}^{\infty} n(n-1)\dots(n-k+1) a_n x^{n-k};$$

in particular, $f^{(k)}(0) = k!\, a_k$.

Proof. This follows at once from 19.7 by an inductive argument.

19.9 COROLLARY. *If the series $\sum_{n=0}^{\infty} b_n p_n$ and $\sum_{n=0}^{\infty} c_n p_n$ converge and have the same sum at each point of some interval $(-r, r)$, $r > 0$, then*

$$b_n = c_n \quad (n = 0, 1, 2, \dots).$$

The reader may well have used 19.8 and 19.9, without proof, to obtain power series solutions of linear differential equations.

Problems on Chapter 3

1. Let (f_n) be the sequence of functions on $\mathbf{R}$ defined by

$$f_n(x) = \begin{cases} \dfrac{n - |x|}{n^2} & \text{if } |x| \leq n \\ 0 & \text{otherwise.} \end{cases}$$

Show that (f_n) converges uniformly on $\mathbf{R}$ to the zero function, but $\lim \int_{-\infty}^{\infty} f_n(x)\,dx \neq 0$. Why does this example not contradict 15.2?

2. Let (f_n) be the sequence of functions on $\mathbf{R}$ defined by

$$f_n(x) = \frac{x}{1 + nx^2} \quad (x \in \mathbf{R};\ n = 1, 2, \dots).$$

Show that (f_n) converges uniformly on **R** to the zero function, but $\lim f_n'(0) \neq 0$. Why does this example not contradict 16.1?

3. Let (f_n) be a sequence of real-valued differentiable functions on an interval (a, b) such that each of the functions f_n' is continuous on (a, b), (f_n') converges uniformly on (a, b) to a function g and there is a point x_0 in (a, b) such that $(f_n(x_0))$ converges. Without using 16.1, show that (f_n) converges uniformly on (a, b) of a function f, f is differentiable on (a, b) and $f' = g$.
[Hint: consider the sequence (F_n) of functions defined by

$$F_n(x) = \int_{x_0}^{x} f_n' \quad (a < x < b; \; n = 1, 2, \ldots).]$$

4. Denote the set of real-valued functions f on $[a, b]$ such that f' is defined and continuous on $[a, b]$ by $C_\mathbf{R}^1[a, b]$ and the uniform norm on $[a, b]$ by $\|\cdot\|$. Show that the normed space $(C_\mathbf{R}^1[a, b], \|\cdot\|)$ is not complete.

5. Define $C_\mathbf{R}^1[a, b]$ as in Problem 4, and write

$$\|f\|_1 = \|f\| + \|f'\| \quad (f \in C_\mathbf{R}^1[a, b]),$$

where $\|\cdot\|$ denotes the uniform norm on $[a, b]$. Show that the normed space $(C_\mathbf{R}^1[a, b], \|\cdot\|_1)$ is complete.

6. Let (f_n) be a sequence of complex-valued functions on $[0, 1)$ such that the series $\sum_{n=1}^{\infty} f_n$ converges uniformly on $[0, 1)$ and, for each positive integer n, $\lim_{x \to 1-} f_n(x) = \alpha_n$, a complex number. Show that the series $\sum_{n=1}^{\infty} \alpha_n$ is convergent and has sum $\lim_{x \to 1-} \left(\sum_{n=1}^{\infty} f_n(x) \right)$.

7. Show that the series $\sum_{n=1}^{\infty} \dfrac{\sin nx}{n}$ does not converge uniformly on $[0, 2\pi]$.

[Hint: $\sin y \geq \dfrac{2y}{\pi}$ for each y in $[0, \tfrac{1}{2}\pi]$.]

8. If $\alpha > \tfrac{1}{2}$, show that the series $\sum_{n=1}^{\infty} \dfrac{x}{n^\alpha(1+nx^2)}$ converges uniformly on **R**.

9. Show that the series $\sum_{n=1}^{\infty} \dfrac{x^n(1-x)}{\log(n+1)}$ converges uniformly on $[0, 1]$, but does not satisfy the conditions of Weierstrass's test.

10. Show that the series $\sum_{n=1}^{\infty} \dfrac{x^n(1-x^n)}{n}$ converges uniformly on $[0, \delta]$ for each δ in $(0, 1)$, but does not converge uniformly on $[0, 1]$.

11. Show that $\lim\limits_{x \to 1} \left(\sum_{n=1}^{\infty} (-1)^{n+1} n^{-x} \right) = \log 2$.

12. Show that $\lim\limits_{x \to 0} \left(\sum_{n=1}^{\infty} \dfrac{1}{(n+x)(n+1)} \right) = 1$.

13. For each pair of positive integers i, j, let a_{ij} be a complex number such that the series $\sum_{j=1}^{\infty} |a_{ij}|$ is convergent with sum M_i and $\sum_{i=1}^{\infty} M_i$ is convergent. Show that

$$\sum_{i=1}^{\infty} \left(\sum_{j=1}^{\infty} a_{ij} \right) = \sum_{j=1}^{\infty} \left(\sum_{i=1}^{\infty} a_{ij} \right).$$

[Hint: consider the functions f_i ($i = 1, 2, \ldots$) and g defined on the set $X = \{1/n : n \text{ a positive integer}\} \cup \{0\}$ by

$$f_i(1/n) = \sum_{j=1}^{n} a_{ij}, \quad f_i(0) = \sum_{j=1}^{\infty} a_{ij}, \quad g(x) = \sum_{i=1}^{\infty} f_i(x).]$$

14. Show that, for each positive real number x,

$$\sum_{j=1}^{\infty} \left(\sum_{i=1}^{\infty} \dfrac{(-1)^{j+1}}{(x+i)^{j+1}} \right) = \dfrac{1}{x+1}.$$

15. Show that the series $\sum_{n=1}^{\infty} x(1-x)^n$ converges for each $x \in [0, 1]$, but does not converge uniformly on this interval. Show, however, that term-by-term integration over $[0, 1]$ is valid. (Thus the converse of 17.4 is false.)

16. Show that the series $\sum_{n=1}^{\infty} \dfrac{1}{n^3 + n^4 x^2}$ converges uniformly on $\mathbf{R}$, the function f defined by

$$f(x) = \sum_{n=1}^{\infty} \frac{1}{n^3 + n^4 x^2} \quad (x \in \mathbf{R})$$

is differentiable everywhere and its derivative can be obtained by term-by-term differentiation.

17. Show that the series $\sum_{n=1}^{\infty} \frac{1}{n^2 + n^4 x^2}$ converges uniformly on $\mathbf{R}$, but the function f defined by

$$f(x) = \sum_{n=1}^{\infty} \frac{1}{n^2 + n^4 x^2} \quad (x \in \mathbf{R})$$

is not differentiable at 0.

[Hint: if $h > 0$, $\sum_{n=1}^{\infty} \frac{h}{1+n^2 h^2} > \int_{1}^{\infty} \frac{h\,dt}{1+h^2 t^2} = \int_{h}^{\infty} \frac{du}{1+u^2}$.]

18. The Riemann zeta-function ζ is defined by

$$\zeta(x) = \sum_{n=1}^{\infty} n^{-x} \quad (x > 1).$$

Show that $\quad \zeta'(x) = -\sum_{n=1}^{\infty} n^{-x} \log n \quad (x > 1).$

19. Let α be a real number, other than a non-negative integer, and write

$$\binom{\alpha}{0} = 1, \quad \binom{\alpha}{n} = \frac{\alpha(\alpha-1)(\alpha-2)\ldots(\alpha-n+1)}{n!} \quad (n = 1, 2, \ldots).$$

Show that the power series $\sum_{n=0}^{\infty} \binom{\alpha}{n} p_n$ has radius of convergence 1. If f is defined by

$$f(x) = \sum_{n=0}^{\infty} \binom{\alpha}{n} x^n \quad (-1 < x < 1),$$

show that $(1+x)f'(x) = \alpha f(x)$ and hence that $f(x) = (1+x)^\alpha$ for each $x \in (-1, 1)$.

20. If (a_n) is a sequence of positive real numbers and

$$\lim \left[n \frac{a_n}{a_{n+1}} - (n+1) \right] = \lambda, \text{ a positive real number,}$$

show that the series $\sum_{n=1}^{\infty} a_n$ is convergent. Hence, or otherwise, show that the series $\sum_{n=1}^{\infty} \binom{\frac{1}{2}}{n}$ is absolutely convergent. Deduce that

(a) $\sum_{n=0}^{\infty} \binom{\frac{1}{2}}{n} x^n$ converges uniformly on $[-1, 0]$ to $\sqrt{(1+x)}$;

(b) if $M, \varepsilon > 0$, there is a polynomial function p such that
$$\sup_{-M \leq y \leq M} \big||y| - p(y)\big| < \varepsilon.$$

$\left[\text{Hint: } |y| = M\left(1 + \frac{y^2 - M^2}{M^2}\right)^{\frac{1}{2}}.\right]$

CHAPTER 4
LEBESGUE INTEGRATION

§20. The collection K and null sets

The integration theory of a first analysis course, namely Riemann integration, has several deficiencies from a theoretical point of view. For one thing, the collection of functions Riemann integrable over an interval $[a, b]$ is not large enough. For instance, we saw in 15.1(a) that there is a sequence (f_n) of bounded functions on $[0, 1]$, converging pointwise on $[0, 1]$ to a bounded function f, such that each function f_n is Riemann integrable over $[0, 1]$, but f is not Riemann integrable over $[0, 1]$. Worse still, it can be shown that the normed space $R[a, b]$ of functions Riemann integrable over $[a, b]$, with norm defined by $\|f\| = \int_a^b |f|$, is not complete. In this chapter, we shall discuss a more general integration theory, formulated by Lebesgue at the beginning of this century. We shall not, however, follow Lebesgue's approach to the subject.

We begin by introducing a collection K of functions.

20.1 DEFINITION. If φ is a complex-valued function on $\mathbf{R}$, the **support** of φ, $S(\varphi)$, is defined by $S(\varphi) = \{x \in \mathbf{R} : \varphi(x) \neq 0\}^-$. Denote by K the collection of continuous real-valued functions on $\mathbf{R}$ having compact support.

Since the support of a complex-valued function on $\mathbf{R}$ is a closed subset of $\mathbf{R}$, it is compact if and only if it is bounded. Thus a continuous real-valued function φ on $\mathbf{R}$ belongs to K if and only if there is a bounded interval $[a, b]$, depending on φ, such that $S(\varphi) \subseteq [a, b]$, i.e. such that $\varphi(x) = 0$ for all $x \in \mathbf{R} - [a, b]$.

If $\varphi \in K$ and $[a, b]$, $[c, d]$ are intervals containing $S(\varphi)$, then

$$\int_a^b \varphi = \int_c^d \varphi.$$

(Here $\int_a^b \varphi$ denotes the Riemann integral of the continuous function φ over the interval $[a, b]$.) For if $\alpha = \min\{a, c\}$ and $\beta = \max\{b, d\}$, then

$$\int_a^b \varphi = \int_\alpha^a \varphi + \int_a^b \varphi + \int_b^\beta \varphi \quad \text{as } \varphi \text{ vanishes on } \mathbf{R} - [a, b]$$
$$= \int_\alpha^\beta \varphi$$

and similarly $\int_c^d \varphi = \int_\alpha^\beta \varphi$. Thus we can make the following definition:

20.2 DEFINITION. If $\varphi \in K$ and $[a, b]$ is any interval containing $S(\varphi)$, we define the **integral** of φ, $I(\varphi)$, by

$$I(\varphi) = \int_a^b \varphi.$$

The following theorem summarizes the important properties of the collection K and the function I on K.

20.3 THEOREM. (a) *K is a real vector space and a lattice, i.e. if $\varphi, \psi \in K$ and $a \in \mathbf{R}$, then $(\varphi + \psi), a\varphi, \varphi \vee \psi, \varphi \wedge \psi \in K$;*
(b) *I is a linear functional on K, i.e. if $\varphi, \psi \in K$ and $a \in \mathbf{R}$, then $I(\varphi + \psi) = I(\varphi) + I(\psi)$ and $I(a\varphi) = aI(\varphi)$;*
(c) *I is increasing, i.e. if $\varphi, \psi \in K$ and $\varphi \geqslant \psi$, then $I(\varphi) \geqslant I(\psi)$;*
(d) *I is continuous under monotone limits, i.e. if (φ_n) is a decreasing sequence of functions in K which converges pointwise on $\mathbf{R}$ to the zero function, then $\lim I(\varphi_n) = 0$.*

Proof. (a) is easily verified. The details are left to the reader. (b) and (c) follow from elementary properties of the Riemann integral. To prove (d), let (φ_n) be a decreasing sequence of functions in K which converges pointwise on $\mathbf{R}$ to the zero function. Let $[a, b]$ be an interval containing $S(\varphi_1)$. Then, since (φ_n) is a decreasing sequence of non-negative functions, we have, for each positive integer n $[a, b] \supseteq S(\varphi_n)$. According to Dini's theorem, 9.7, the sequence (φ_n) converges uniformly on $[a, b]$ to the zero function. Hence, by 15.2,

$$\lim I(\varphi_n) = \lim \int_a^b \varphi_n = \int_a^b (\lim \varphi_n) = 0.$$

It is easily deduced from this theorem that if (ψ_n) is a monotonic sequence of functions in K which converges pointwise on $\mathbf{R}$ to a function φ in K, then $\lim I(\psi_n) = I(\varphi)$. This explains the terminology in 20.3(d).

It is our plan to use a limiting procedure to extend the integral I, defined on K, to an integral, which we shall also denote by I, on a larger collection of functions, the Lebesgue integrable functions. The above comment suggests a possible first step: namely, if f is a real-valued function on $\mathbf{R}$ for which there is, say, an increasing sequence (φ_n) of functions in K such that $(I(\varphi_n))$ is bounded and

$$f(x) = \lim \varphi_n(x) \text{ for all } x \text{ in } \mathbf{R}, \tag{20.1}$$

define $I(f) = \lim I(\varphi_n)$. (This equation *is* correct if $f \in K$.)

A moment's thought, however, suggests that (20.1) is perhaps an unnecessarily restrictive condition. If f is Riemann integrable over an interval $[a, b]$ and g is a function on $[a, b]$ such that $g(x) = f(x)$ except at a finite number of points of $[a, b]$, then g is Riemann integrable over $[a, b]$ and $\int_a^b g = \int_a^b f$. In other words, finite sets are negligible as far as Riemann integration is concerned. It is only reasonable to expect there to be subsets of $\mathbf{R}$ which are negligible with respect to Lebesgue integration. We now define these sets.

20.4 DEFINITION. A subset E of $\mathbf{R}$ is said to be **null** if there is an increasing sequence (φ_n) of non-negative functions in K such that $\lim \varphi_n(x) = \infty$ for each $x \in E$ but $\sup I(\varphi_n) \, (= \sup_{n \geq 1} I(\varphi_n)) < \infty$.

It is clear from this definition that if a subset E of $\mathbf{R}$ is null, then so is any subset of E. Our next result describes a more important property of null sets.

20.5 THEOREM. *Let (E_n) be a sequence of null sets. Then the set $\bigcup_{n=1}^{\infty} E_n$ is also null.*

Proof. For each positive integer k, let (φ_{kn}) be an increasing sequence of non-negative functions in K such that $\lim_{n \to \infty} \varphi_{kn}(x) = \infty$

for each $x \in E_k$ and $\sup I(\varphi_{kn}) < \infty$. Multiplying each of the functions φ_{kn}, $n = 1, 2, \ldots$, by a suitable positive real number, we may assume that $\sup I(\varphi_{kn}) \leq 2^{-k}$. Write

$$\varphi_n = \varphi_{1n} \vee \varphi_{2n} \vee \ldots \vee \varphi_{nn} \quad (n = 1, 2, \ldots).$$

Then (φ_n) is an increasing sequence of non-negative functions in K and, since $\varphi_n \geq \varphi_{kn}$ for all $n \geq k$, we have

$$\lim \varphi_n(x) = \infty \text{ for each } x \in \bigcup_{n=1}^{\infty} E_n.$$

Finally, since $\varphi_n \leq \sum_{k=1}^{n} \varphi_{kn}$,

$$I(\varphi_n) \leq \sum_{k=1}^{n} I(\varphi_{kn}) < \sum_{k=1}^{n} 2^{-k} < 1.$$

Thus $\bigcup_{n=1}^{\infty} E_n$ is a null set.

Although 20.4 is the most convenient definition of null set for our purposes, it is hardly illuminating. It does not make it at all clear what null sets are like. We shall therefore establish alternative characterizations of these sets. First we define the length of an open subset of **R**.

Let G be an open subset of **R**. Then, by 4.9, G is the union of a countable collection $(a_1, b_1), (a_2, b_2), \ldots$ of disjoint open intervals. We define the **length** of G, $l(G)$, by

$$l(G) = \sum_{n \geq 1} (b_n - a_n).$$

We shall require the following result about lengths of open sets.

20.6 LEMMA. *Let (G_n) be a sequence of open subsets of* **R** *such that $G_{n+1} \supseteq G_n$ for each positive integer n and $\sup l(G_n) < \infty$: write $G = \bigcup_{n=1}^{\infty} G_n$. Then*

$$l(G) = \sup l(G_n).$$

Proof. Write $\alpha = \sup l(G_n)$. Clearly $l(G_n) \leq l(G)$ $(n = 1, 2, \ldots)$, so that $\alpha \leq l(G)$. To prove the reverse inequality, we argue by contradiction. Suppose $l(G) > \alpha$, say $l(G) = \alpha + 3\varepsilon$ where $\varepsilon > 0$. G, being an open subset of **R**, is the union of a countable collection $(a_1, b_1), (a_2, b_2), \ldots$ of disjoint open intervals. Since $l(G) = \sum_{n \geq 1} (b_n - a_n) = \alpha + 3\varepsilon$, there is a positive integer p such that

$\sum_{n=1}^{p} (b_n - a_n) > \alpha + 2\varepsilon$. Let δ be a positive real number less than $\min \{\frac{1}{2}(b_1 - a_1), \ldots, \frac{1}{2}(b_p - a_p), \varepsilon/p\}$, and consider the intervals $F_n = [a_n + \delta, b_n - \delta]$, $n = 1, 2, \ldots, p$. Each of these intervals is compact and is contained in $\bigcup_{n=1}^{\infty} G_n$. Hence, since $G_{n+1} \supseteq G_n$ ($n = 1, 2, \ldots$), there is, for $n = 1, 2, \ldots, p$, a positive integer m_n such that $G_{m_n} \supseteq F_n$. Let $M = \max \{m_1, m_2, \ldots, m_p\}$. Then, for $n = 1, 2, \ldots, p$, we have $G_M \supseteq G_{m_n} \supseteq F_n$. Since the intervals $F_1, F_2, \ldots, F_p$ are disjoint, it follows that

$$l(G_M) \geq \sum_{n=1}^{p} (b_n - a_n - 2\delta)$$
$$> \sum_{n=1}^{p} (b_n - a_n) - 2\varepsilon \quad (-2p\delta > -2\varepsilon)$$
$$> \alpha,$$

contradicting the definition of α.

We are now in a position to establish the alternative characterizations of null sets.

20.7 THEOREM. *Let E be a subset of $\mathbf{R}$. Then the following statements are equivalent:*

(a) *for each $\varepsilon > 0$, there is a countable collection $I_1, I_2, \ldots$ of open intervals such that*

$$\bigcup_{n \geq 1} I_n \supseteq E \quad \text{and} \quad \sum_{n \geq 1} l(I_n) < \varepsilon;$$

(b) *there is a countable collection $I_1, I_2, \ldots$ of open intervals such that each point of E belongs to infinitely many of the intervals I_n and $\sum_{n \geq 1} l(I_n) < \infty$;*

(c) *E is null.*

Proof. We prove the result by showing that (a) $\Rightarrow$ (b) $\Rightarrow$ (c) $\Rightarrow$ (a).

Suppose first that (a) is satisfied. Then, for each positive integer k, there is a countable collection $I_{k1}, I_{k2}, \ldots$ of open intervals such that

$$\bigcup_{n \geq 1} I_{kn} \supseteq E \quad \text{and} \quad \sum_{n \geq 1} l(I_{kn}) < 2^{-k}.$$

$\{I_{kn} : k, n = 1, 2, \ldots\}$ is a countable collection of open intervals; for each $x \in E$ and each positive integer k, there is a positive integer n

D

such that $x \in I_{kn}$; further $\sum_{k,n \geq 1} l(I_{kn}) < \sum_{k=1}^{\infty} 2^{-k} = 1$. Thus (b) is satisfied.

Suppose now that (b) is satisfied, and let $I_1, I_2, \ldots$ be a countable collection of open intervals such that each point of E belongs to infinitely many of the intervals I_n and $\sum_{n \geq 1} l(I_n) < \infty$. Suppose that $I_n = (a_n, b_n), n = 1, 2, \ldots$. For each positive integer k, let ψ_k be the function in K having the graph shown in Figure 20.1, and write $\varphi_n = \sum_{k=1}^{n} \psi_k$ $(n = 1, 2, \ldots)$. Then (φ_n) is an increasing sequence of non-negative functions in K and

$$I(\varphi_n) = \sum_{k=1}^{n} I(\psi_k) = \sum_{k=1}^{n} (b_k - a_k + 2^{-k}) < \sum_{k \geq 1} l(I_k) + 1.$$

Further if $x \in E$, then $a_k < x < b_k$ for infinitely many k; for such k, $\psi_k(x) = 1$. Hence $\lim \varphi_n(x) = \infty$ for each $x \in E$. Thus E is a null set and (c) is satisfied.

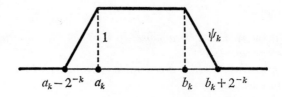

Figure 20.1. The graph of ψ_k.

Finally, suppose that E is a null set, and let (φ_n) be an increasing sequence of non-negative functions in K such that $\lim \varphi_n(x) = \infty$ for each $x \in E$ and $\sup I(\varphi_n) = M$, a real number. Let $\varepsilon > 0$. Then

$$E \subseteq \{x \in \mathbf{R} : \varphi_n(x) > 2M/\varepsilon \text{ for some } n\}$$

$$= \bigcup_{n=1}^{\infty} \{x \in \mathbf{R} : \varphi_n(x) > 2M/\varepsilon\}. \quad (20.2)$$

For each positive integer n, the set $G_n = \{x \in \mathbf{R} : \varphi_n(x) > 2M/\varepsilon\}$ is open as φ_n is continuous, and $G_{n+1} \supseteq G_n$ as $\varphi_{n+1} \geq \varphi_n$. Further, since φ_n is non-negative, $M \geq I(\varphi_n) > \dfrac{2M}{\varepsilon} l(G_n)$, so that $l(G_n) < \tfrac{1}{2}\varepsilon$.

$G = \bigcup_{n=1}^{\infty} G_n$ is an open subset of **R** and, by 20.6, $I(G) \leq \frac{1}{2}\varepsilon$. The inclusion (20.2) therefore shows that (a) is satisfied.

(a) gives perhaps the most understandable characterization of a null set. For example, it makes it clear that every finite subset of **R** is null.

20.8 DEFINITION. If a property holds for all real numbers outside some null set, we say that the property holds **almost everywhere** (a.e.) or that it holds for **almost all** (a.a.) points of **R**.

For example, if (φ_n) is an increasing sequence of non-negative functions in K and $\sup I(\varphi_n) < \infty$, then, by the very definition of a null set, (φ_n) converges a.e. or equivalently $(\varphi_n(x))$ converges for a.a. x in **R**.

§21. The Lebesgue integral

21.1 DEFINITION. Denote by K^+ the collection of all real-valued functions f on **R** for which there is an increasing sequence (φ_n) of functions in K such that

$$\sup I(\varphi_n) < \infty \text{ and } f(x) = \lim \varphi_n(x) \text{ for a.a. } x \text{ in } \mathbf{R}. \quad (21.1)$$

Clearly $K^+ \supseteq K$. In fact, K is a proper subset of K^+. For example, if $a < b$, the function f on **R** defined by

$$f(x) = \begin{cases} 1 & \text{if } a \leq x \leq b \\ 0 & \text{otherwise} \end{cases}$$

is not in K but does belong to K^+. If φ_n is the function having the graph shown in Figure 21.1, then (φ_n) is an increasing sequence of functions in K, $\sup I(\varphi_n) = (b-a)$ and $\lim \varphi_n(x) = f(x)$ for all $x \in \mathbf{R} - \{a, b\}$. The function f is called the **characteristic function** of the interval $[a, b]$, and is often denoted by $\chi_{[a, b]}$.

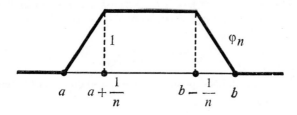

Figure 21.1. The graph of φ_n.

If $f \in K^+$ and (φ_n) is an increasing sequence of functions in K satisfying (21.1), it seems natural to define

$$I(f) = \lim I(\varphi_n), \qquad (21.2)$$

thus extending I to K^+. $((I(\varphi_n))$ is a bounded, increasing sequence, and so certainly converges.) There is, however, one obvious objection to doing this—the right-hand side of (21.2) might depend on the choice of the increasing sequence (φ_n) of functions in K satisfying (21.1). We shall show presently that this is not the case. First we must improve on the result given in 20.3(d).

21.2 LEMMA. *Let (φ_n) be a decreasing sequence of non-negative functions in K which converges to the zero function a.e. Then $\lim I(\varphi_n) = 0$.*

Proof. Write $E = \{x \in \mathbf{R}: \lim \varphi_n(x) \neq 0\}$. By hypothesis, E is a null set, and so there is an increasing sequence (ψ_n) of non-negative functions in K such that $\lim \psi_n(x) = \infty$ for each $x \in E$ and $\lim I(\psi_n) = M$, a real number. Let $\varepsilon > 0$, and define

$$\theta_n = (\varphi_n - \varepsilon\psi_n) \vee 0 \quad (n = 1, 2, \ldots).$$

Then (θ_n) is a decreasing sequence of non-negative functions in K and $\lim \theta_n(x) = 0$ for all $x \in \mathbf{R}$. Hence, by 20.3(d),

$$\lim I(\theta_n) = 0.$$

Now, for each positive integer n,

$$0 \leq \varphi_n = (\varphi_n - \varepsilon\psi_n) + \varepsilon\psi_n \leq \theta_n + \varepsilon\psi_n,$$

so that

$$0 \leq \lim I(\varphi_n) \leq \lim I(\theta_n) + \lim I(\varepsilon\psi_n) = \varepsilon M.$$

Since $\varepsilon > 0$ is arbitrary, the result follows.

21.3 THEOREM. *Let (φ_n), (ψ_n) be increasing sequences of functions in K such that $\sup I(\varphi_n) < \infty$, $\sup I(\psi_n) < \infty$ and $\lim \varphi_n(x) \geq \lim \psi_n(x)$ for a.a. x in $\mathbf{R}$. Then*

$$\lim I(\varphi_n) \geq \lim I(\psi_n).$$

Proof. Let m be a fixed positive integer, and consider the sequence (θ_n) defined by

$$\theta_n = (\psi_m - \varphi_n) \vee 0 \quad (n = 1, 2, \ldots).$$

Then (θ_n) is a decreasing sequence of non-negative functions in K, and $\lim \theta_n(x) = 0$ for a.a. x in $\mathbf{R}$. Hence, by 21.2,
$$\lim I(\theta_n) = 0.$$
Now, for each positive integer n, $(\psi_m - \varphi_n) \leqslant \theta_n$, and so
$$I(\psi_m) - \lim I(\varphi_n) = \lim I(\psi_m - \varphi_n) \leqslant \lim I(\theta_n) = 0$$
i.e.
$$I(\psi_m) \leqslant \lim I(\varphi_n).$$
Since this is true for each positive integer m, the result follows.

21.4 COROLLARY. *If (φ_n), (ψ_n) are increasing sequences of functions in K such that $\sup I(\varphi_n) < \infty$, $\sup I(\psi_n) < \infty$ and*
$$\lim \varphi_n(x) = \lim \psi_n(x) \text{ for a.a. } x \text{ in } \mathbf{R}, \tag{21.3}$$
then
$$\lim I(\varphi_n) = \lim I(\psi_n).$$

Proof. (21.3) is equivalent to the two inequalities
$$\lim \varphi_n(x) \geqslant \lim \psi_n(x) \text{ for a.a. } x \text{ in } \mathbf{R}$$
and
$$\lim \psi_n(x) \geqslant \lim \varphi_n(x) \text{ for a.a. } x \text{ in } \mathbf{R}.$$

21.5 DEFINITION. If $f \in K^+$ and (φ_n) is any increasing sequence of functions in K satisfying (21.1), we define the integral of f, $I(f)$, by
$$I(f) = \lim I(\varphi_n).$$
21.4 shows that $I(f)$ is well-defined and that if $f \in K$, the above definition agrees with the original one.

We note that if f, g are real-valued functions on $\mathbf{R}$, $f \in K^+$ and $g = f$ a.e., then
$$g \in K^+ \text{ and } I(g) = I(f). \tag{21.4}$$

We now show that the collection K^+ and the integral on K^+ possess some of the properties of K and the integral on K mentioned in 20.3.

21.6 THEOREM. (a) *Let $f, g \in K^+$ and a be a non-negative real number. Then $(f+g)$, af, $f \vee g$, $f \wedge g \in K^+$ and $I(f+g) = I(f) + I(g)$, $I(af) = aI(f)$.*
(b) *If $f, g \in K^+$ and $f \geqslant g$, then $I(f) \geqslant I(g)$.*

Proof. (a) Since $f, g \in K^+$, there are increasing sequences (φ_n), (ψ_n) of functions in K such that sup $I(\varphi_n) < \infty$, sup $I(\psi_n) < \infty$ and the sets

$$A = \{x \in \mathbf{R}: \lim \varphi_n(x) \neq f(x)\}, \quad B = \{x \in \mathbf{R}: \lim \psi_n(x) \neq g(x)\}$$

are both null. $(\varphi_n + \psi_n)$ is an increasing sequence of functions in K, sup $I(\varphi_n + \psi_n) < \infty$ and

$$\lim (\varphi_n(x) + \psi_n(x)) = f(x) + g(x) \text{ for all } x \in \mathbf{R} - (A \cup B).$$

By 20.5, $A \cup B$ is a null set. Hence $(f+g) \in K^+$ and

$$I(f+g) = \lim I(\varphi_n + \psi_n) = \lim I(\varphi_n) + \lim I(\psi_n) = I(f) + I(g).$$

The remaining results in (a) are proved in a similar way. The details are left to the reader.

(b) is simply a restatement of 21.3.

We note that K^+ is not closed under multiplication by negative real numbers. For example, the function f defined on $\mathbf{R}$ by

$$f(x) = \begin{cases} \dfrac{1}{\sqrt{x}} & \text{if } 0 < x < 1 \\ 0 & \text{otherwise} \end{cases}$$

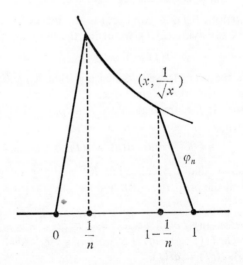

Figure 21.2

belongs to K^+. If φ_n is the function having the graph shown in Figure 21.2, then (φ_n) is an increasing sequence of functions in K, $\lim \varphi_n(x) = f(x)$ for all $x \in \mathbf{R}$, and $I(\varphi_n)$ is less than the improper Riemann integral $\int_0^1 \frac{dx}{\sqrt{x}}$. However $-f \notin K^+$, for clearly there is no function φ in K such that $\varphi \leq -f$ a.e. Thus K^+ is not a vector space.

The collection of functions which interests us, namely the Lebesgue integrable functions, turns out to be the smallest vector space of functions containing K^+.

21.7 DEFINITION. Denote by $L^1 = L^1(\mathbf{R})$ the collection of all functions of the form $(f_1 - f_2)$ with $f_1, f_2 \in K^+$. The functions in L^1 are said to be **(Lebesgue) integrable**.

Clearly $L^1 \supseteq K^+$. In fact, the above discussion shows that the inclusion is proper.

Let $f \in L^1$, say $f = f_1 - f_2$ where $f_1, f_2 \in K^+$. Then it is obvious that we should define $I(f) = I(f_1) - I(f_2)$, thus extending I to L^1. Once again, however, we must show that such a definition is possible. Suppose we also have $f = g_1 - g_2$ where $g_1, g_2 \in K^+$. Then $f_1 + g_2 = f_2 + g_1$, and so, by 21.6(a), $I(f_1) + I(g_2) = I(f_2) + I(g_1)$, i.e.

$$I(f_1) - I(f_2) = I(g_1) - I(g_2). \tag{21.5}$$

21.8 DEFINITION. If $f \in L^1$, say $f = f_1 - f_2$ where $f_1, f_2 \in K^+$ then we define $I(f) = I(f_1) - I(f_2)$. The function I, so defined on L^1, is called the **Lebesgue integral**. (We shall occasionally find it convenient to denote the Lebesgue integral of f by $\int_{-\infty}^{\infty} f(x)\,dx$.)

(21.5) shows that $I(f)$ is well-defined and that if $f \in K^+$, the above definition agrees with the previous one.

We note that if f, g are real-valued functions on $\mathbf{R}$, $f \in L^1$ and $g = f$ a.e., then, as is easily shown using (21.4), $g \in L^1$ and $I(g) = I(f)$. Thus null sets are negligible as far as Lebesgue integration is concerned.

We conclude this section by establishing some of the elementary properties of L^1 and the Lebesgue integral.

21.9 THEOREM. (a) *L^1 is a real vector space and a lattice;* (b) *the Lebesgue integral is an increasing linear functional on L^1.*

Proof. Let $f, g \in L^1$, say $f = f_1 - f_2$ and $g = g_1 - g_2$ where $f_1, f_2, g_1, g_2 \in K^+$, and $a \in \mathbf{R}$. Then, by 21.6(a),

$$f + g = (f_1 + g_1) - (f_2 + g_2) \in L^1,$$

and

$$\begin{aligned} I(f+g) &= I(f_1 + g_1) - I(f_2 + g_2) \\ &= I(f_1) + I(g_1) - I(f_2) - I(g_2) \\ &= I(f_1) - I(f_2) + I(g_1) - I(g_2) \\ &= I(f) + I(g). \end{aligned}$$

If $a \geqslant 0$, then, by 21.6(a) again,

$$af = af_1 - af_2 \in L^1$$

and

$$I(af) = I(af_1) - I(af_2) = aI(f_1) - aI(f_2) = aI(f);$$

if $a < 0$, then

$$af = (-a)f_2 - (-a)f_1 \in L^1$$

and

$$I(af) = I((-a)f_2) - I((-a)f_1) = (-a)I(f_2) - (-a)I(f_1) = aI(f).$$

Thus L^1 is a vector space and I is a linear functional on L^1.

To show that L^1 is a lattice, we first note that

$$|f| = (f_1 \vee f_2) - (f_1 \wedge f_2) \in L^1.$$

Hence

$$f \vee g = \tfrac{1}{2}(f + g + |f - g|) \quad \text{and} \quad f \wedge g = \tfrac{1}{2}(f + g - |f - g|)$$

belong to L^1.

To show that I is increasing on L^1, suppose that $f \geqslant g$. Then $f_1 + g_2 \geqslant f_2 + g_1$, so that, by 21.6(b), $I(f_1) + I(g_2) \geqslant I(f_2) + I(g_1)$ i.e.

$$I(f) = I(f_1) - I(f_2) \geqslant I(g_1) - I(g_2) = I(g).$$

21.10 COROLLARY. *If $f \in L^1$, then $|I(f)| \leqslant I(|f|)$.*

Proof. $-|f| \leqslant f \leqslant |f|$, so that $-I(|f|) \leqslant I(f) \leqslant I(|f|)$.

Let $f \in L^1$. Then, by definition, f has a decomposition $f = f_1 - f_2$ with $f_1, f_2 \in K^+$. It is worth pointing out that, in this decomposition, we may assume that f_2 is non-negative and $I(f_2)$ is arbitrarily small.

21.11 Theorem. *Let $f \in L^1$ and $\varepsilon > 0$. Then there are functions g, h in K^+ such that $f = g - h$, h is non-negative and $I(h) < \varepsilon$.*

Proof. Let f_1, f_2 be functions in K^+ such that $f = f_1 - f_2$. There is an increasing sequence (φ_n) of functions in K such that the set $E = \{x \in \mathbf{R} : f_2(x) \neq \lim \varphi_n(x)\}$ is null. By definition, we have $I(f_2) = \lim I(\varphi_n)$, and so we can choose a positive integer n_0 such that $I(f_2) < I(\varphi_{n_0}) + \varepsilon$. Define the functions g, h on $\mathbf{R}$ by

$$g(x) = \begin{cases} f_1(x) - \varphi_{n_0}(x) & (x \notin E) \\ f(x) & (x \in E) \end{cases}, \quad h(x) = \begin{cases} f_2(x) - \varphi_{n_0}(x) & (x \notin E) \\ 0 & (x \in E) \end{cases}.$$

Then $g, h \in K^+$, $f = g - h$, h is non-negative and

$$I(h) = I(f_2) - I(\varphi_{n_0}) < \varepsilon.$$

§22. Convergence theorems

In this section, we shall establish two important theorems giving conditions under which the limit function f of a convergent sequence (f_n) of L^1 functions is in L^1 and the relation $I(f) = \lim I(f_n)$ holds. To prove the first of these theorems, in which the sequence (f_n) is assumed to be monotonic, we require a preliminary lemma.

22.1 Lemma. *Let (f_n) be an increasing sequence of functions in K^+ such that $\sup I(f_n) < \infty$. Then there is a function f in K^+ such that*

$$f(x) = \lim f_n(x) \quad \text{for a.a. } x \text{ in } \mathbf{R};$$

further

$$I(f) = \lim I(f_n).$$

(An alternative way of stating the conclusion is: the sequence (f_n) converges a.e. and if f is any real-valued function on $\mathbf{R}$ such that $f(x) = \lim f_n(x)$ for a.a. x in $\mathbf{R}$, then $f \in K^+$ and $I(f) = \lim I(f_n)$.)

Proof. For each positive integer n, let $\varphi_{n1}, \varphi_{n2}, \ldots$ be an increasing sequence of functions in K such that

$$E_n = \{x \in \mathbf{R} : \lim_{k \to \infty} \varphi_{nk}(x) \neq f_n(x)\}$$

is a null set, and write

$$\psi_k = \varphi_{1k} \vee \varphi_{2k} \vee \ldots \vee \varphi_{kk} \quad (k = 1, 2, \ldots).$$

Then (ψ_k) is an increasing sequence of functions in K, and, for each positive integer k,

$$\varphi_{nk}(x) \leq f_n(x) \leq f_k(x) \quad \text{for all } x \in \mathbf{R} - E_n \text{ and } n \leq k,$$

so that

$$\psi_k(x) \leq f_k(x) \text{ for all } x \in \mathbf{R} - \bigcup_{n=1}^{k} E_n.$$

Since $\bigcup_{n=1}^{k} E_n$ is a null set, it follows that

$$I(\psi_k) \leq I(f_k) \leq \sup I(f_n) \quad (k = 1, 2, \ldots).$$

Hence there is a function f in K^+ such that

$$E = \{x \in \mathbf{R} : \lim_{k \to \infty} \psi_k(x) \neq f(x)\}$$

is a null set. Since $\varphi_{nk} \leq \psi_k$ for all $n \leq k$, we have, letting $k \to \infty$,

$$f_n(x) \leq f(x) \text{ for all } x \in \mathbf{R} - (E \cup E_n).$$

If $x \in \mathbf{R} - (E \cup \bigcup_{n=1}^{\infty} E_n)$, $f_n(x)$ is sandwiched between $f(x)$ and $\psi_n(x)$ and $\lim_{n \to \infty} \psi_n(x) = f(x)$. Thus (f_n) converges to f a.e. and

$$I(f) = \lim I(\psi_n) \leq \lim I(f_n) \leq I(f), \text{ i.e. } I(f) = \lim I(f_n).$$

22.2 MONOTONE CONVERGENCE THEOREM. *Let (f_n) be a monotonic sequence of functions in L^1 such that the sequence $(I(f_n))$ is bounded. Then there is a function f in L^1 such that*

$$f(x) = \lim f_n(x) \text{ for a.a. } x \text{ in } \mathbf{R};$$

further

$$I(f) = \lim I(f_n).$$

Proof. We shall prove the result in the case in which the sequence (f_n) is increasing.

For each positive integer n,

$$f_n = \sum_{i=1}^{n-1} (f_{i+1} - f_i) + f_1.$$

By 21.11, we can write $f_{i+1} - f_i = g_i - h_i$ where g_i, h_i are non-negative functions in K^+ and $I(h_i) < 2^{-i}$ $(i = 1, 2, \ldots)$. Then

$$f_n = \sum_{i=1}^{n-1} (g_i - h_i) + f_1 = G_n - H_n + f_1$$

where

$$G_n = \sum_{i=1}^{n-1} g_i \quad \text{and} \quad H_n = \sum_{i=1}^{n-1} h_i.$$

Now (G_n), (H_n) are increasing sequences of functions in K^+,

$$I(H_n) = \sum_{i=1}^{n-1} I(h_i) < \sum_{i=1}^{n-1} 2^{-i} < 1$$

and

$$I(G_n) = I(f_n) + I(H_n) - I(f_1)$$
$$< \sup I(f_n) + 1 - I(f_1).$$

Hence, by 22.1, there are functions G, H in K^+ such that

$$G(x) = \lim G_n(x), \ H(x) = \lim H_n(x) \text{ for a.a. } x \text{ in } \mathbf{R}.$$

The function $f = G - H + f_1$ belongs to L^1 and for a.a. x in $\mathbf{R}$ we have

$$f(x) = \lim G_n(x) - \lim H_n(x) + f_1(x) = \lim f_n(x).$$

Finally

$$\begin{aligned} I(f) &= I(G) - I(H) + I(f_1) \\ &= \lim I(G_n) - \lim I(H_n) + I(f_1) \quad \text{by 22.1} \\ &= \lim I(G_n - H_n + f_1) \\ &= \lim I(f_n). \end{aligned}$$

22.3 COROLLARY. *The Lebesgue integral is continuous under monotone limits.*

(The monotone convergence theorem, in fact, asserts rather more than this.)

We have already pointed out that if $f = 0$ a.e., then $I(f) = 0$. 22.2 implies a partial converse.

22.4 COROLLARY. *If f is a non-negative function in L^1 and $I(f) = 0$ then $f = 0$ a.e.*

Proof. Apply 22.2 to the sequence (nf).

22.2 shows that if (f_n) is a monotonic sequence of functions in L^1, then $\lim I(f_n) = I(\lim f_n)$, provided only that $\lim I(f_n)$ is finite. We should like to drop the condition that the sequence (f_n) is monotonic.

and require only that the sequence (f_n) converges a.e. We note, however, that some additional restriction must be imposed on the sequence (f_n) to ensure any worthwhile conclusion. For the function f_n defined on **R** by

$$f_n(x) = \begin{cases} \dfrac{n-|x|}{n^2} & \text{if } |x| \leqslant n \\ 0 & \text{otherwise} \end{cases}$$

belongs to K, the sequence (f_n) converges on **R** to the zero function but $\lim I(f_n) \neq 0$. (See Problem 3.1.) Such an additional restriction was discovered by Lebesgue. We shall deduce Lebesgue's theorem from the following lemma, which is itself of some importance.

22.5 Fatou's Lemma. *Let (f_n) be a sequence of non-negative functions in L^1 such that $\liminf I(f_n) < \infty$. Then there is a function f in L^1 such that*

$$f(x) = \liminf f_n(x) \text{ for a.a. } x \text{ in } \mathbf{R};$$

further

$$I(f) \leqslant \liminf I(f_n).$$

(Note: the condition that $\liminf I(f_n) < \infty$ is certainly satisfied if $\sup I(f_n) < \infty$.)

Proof. For each positive integer n, define the function g_n by

$$g_n(x) = \inf_{i \geqslant n} f_i(x) \quad (x \in \mathbf{R}).$$

Then $g_n = \lim_{j \to \infty} (f_n \wedge f_{n+1} \wedge \ldots \wedge f_{n+j})$ and, since $(f_n \wedge f_{n+1} \wedge \ldots \wedge f_{n+j})_j$ is a decreasing sequence of non-negative functions in L^1, 22.2 implies that $g_n \in L^1$. The sequence (g_n) is increasing and

$$I(g_n) \leqslant \inf_{i \geqslant n} I(f_i) \leqslant \liminf I(f_n) < \infty.$$

Hence, by 22.2 again, there is a function f in L^1 such that

$$f(x) = \lim g_n(x) = \liminf f_n(x) \text{ for a.a. } x \text{ in } \mathbf{R}$$

and

$$I(f) = \lim I(g_n) \leqslant \liminf I(f_n).$$

22.6 Dominated Convergence Theorem. *Let (f_n) be a sequence of functions in L^1 which converges a.e. to a function f, and suppose that there is a function g in L^1 such that $|f_n| \leq g$ for each positive integer n. Then $f \in L^1$ and*

$$I(f) = \lim I(f_n).$$

(Note: since the sequence (f_n) is not assumed to be monotonic, the convergence of the sequence $(I(f_n))$ is a non-trivial part of the result.)

Proof. $(g-f_n)$ is a sequence of non-negative functions in L^1 and $g-f_n \leq 2g$ $(n = 1, 2, \ldots)$, so that $\liminf I(g-f_n) \leq 2I(g)$. Hence, by 22.5, $g - f$, which equals $\liminf (g-f_n)$ a.e., belongs to L^1 and

$$I(g) - I(f) \leq \liminf I(g-f_n) = I(g) - \limsup I(f_n),$$

i.e.

$$I(f) \geq \limsup I(f_n). \tag{22.1}$$

Similarly we may apply 22.5 to the sequence $(g+f_n)$ which converges a.e. to $(g+f)$. We obtain the inequality

$$I(f) \leq \liminf I(f_n). \tag{22.2}$$

(22.1) and (22.2) together give the result.

22.7 Corollary. *Let (f_n) be a sequence of functions in L^1 which converges a.e. to a function f, and suppose that there is a function g in L^1 such that $|f| \leq g$. Then $f \in L^1$.*

Proof. For each positive n, let

$$g_n = g \wedge (f_n \vee (-g)).$$

Then (g_n) is a sequence of functions in L^1, converging a.e. to f, and $|g_n| \leq g$ $(n = 1, 2, \ldots)$. Hence, by 22.6, $f \in L^1$.

We note that in 22.7 we do *not* assert that $I(f) = \lim I(f_n)$. The result is, however, very useful in showing that a function belongs to L^1.

In many calculations, it is important to know that it is permissible to interchange the order of an integration and a limit operation. The monotone and dominated convergence theorems can often be used to justify such a procedure. A fairly typical argument is illustrated in the proof of the following result.

22.8 Theorem. *Let f be a real-valued function on a strip of the form $J \times \mathbf{R}$ where J is an open interval in $\mathbf{R}$. Suppose that $f(x, \cdot) \in L^1$ for each $x \in J$, and define Φ on J by*

$$\Phi(x) = I(f(x, \cdot)) = \int_{-\infty}^{\infty} f(x, y)\, dy \qquad (x \in J).$$

Suppose further that $D_1 f$ is defined at each point of $J \times \mathbf{R}$ and there is a function g in L^1 such that

$$|D_1 f(x, y)| \leqslant g(y) \text{ for all } (x, y) \in J \times \mathbf{R}.$$

Then, for each $x \in J$, $D_1 f(x, \cdot) \in L^1$, Φ is differentiable at x and

$$\Phi'(x) = I(D_1 f(x, \cdot)) = \int_{-\infty}^{\infty} D_1 f(x, y)\, dy. \tag{22.3}$$

(Thus the derivative of Φ can be obtained by differentiating under the integral sign.)

Proof. Let $x \in J$ and (h_n) be a sequence of non-zero real numbers such that $(x + h_n) \in J$ and $\lim h_n = 0$. Then, for each positive integer n,

$$\frac{\Phi(x + h_n) - \Phi(x)}{h_n} = \int_{-\infty}^{\infty} \frac{f(x + h_n, y) - f(x, y)}{h_n}\, dy.$$

Since $g \in L^1$ and, for each $y \in R$, we have

$$\left| \frac{f(x + h_n, y) - f(x, y)}{h_n} \right| = |D_1 f(x + \theta h_n, y)| \quad \text{for some } \theta \text{ in } (0, 1)$$

$$\leqslant g(y),$$

$$\lim_{n \to \infty} \frac{f(x + h_n, y) - f(x, y)}{h_n} = D_1 f(x, y),$$

the dominated convergence theorem implies that $D_1 f(x, \cdot) \in L^1$ and

$$\lim_{n \to \infty} \frac{\Phi(x + h_n) - \Phi(x)}{h_n} = \int_{-\infty}^{\infty} D_1 f(x, y)\, dy$$

Since this is true for every appropriate sequence (h_n) which converges to 0, Φ is differentiable at x and $\Phi'(x)$ is given by (22.3).

§23. Relation between Riemann and Lebesgue integration

In §20, we asserted that Lebesgue's integration theory is more general than the Riemann theory. In this section, we shall see in what sense this is true.

Let f be a real-valued function on an interval $[a, b]$. Then f is said to be **Lebesgue integrable over** $[a, b]$ if the function g on $\mathbf{R}$ defined by

$$g(x) = \begin{cases} f(x) & \text{if } x \in [a, b] \\ 0 & \text{otherwise} \end{cases} \quad (23.1)$$

belongs to L^1. In this case, we define the Lebesgue integral of f over $[a, b]$, $L\int_a^b f$, by

$$L\int_a^b f = I(g).$$

If f is bounded and Riemann integrable over $[a, b]$, we shall, for the moment, denote the Riemann integral of f over $[a, b]$ by $R\int_a^b f$.

23.1 Theorem. *Let f be a bounded, real-valued function on an interval $[a, b]$, and suppose that f is Riemann integrable over $[a, b]$. Then f is Lebesgue integrable over $[a, b]$ and*

$$L\int_a^b f = R\int_a^b f.$$

Proof. Let g be the function defined in (23.1). With each partition $P: a = x_0 < x_1 < \ldots < x_n = b$ of $[a, b]$, we associate the functions φ, ψ on $\mathbf{R}$ defined by

$$\varphi(x) = 0 = \psi(x) \text{ if } x \in \mathbf{R} - [a, b], \; \varphi(b) = f(b) = \psi(b),$$

$$\varphi(x) = M_i, \; \psi(x) = m_i \text{ if } x_{i-1} \leqslant x < x_i \; (i = 1, 2, \ldots, n),$$

where M_i, m_i denote the supremum and infimum respectively of $\{f(t): x_{i-1} \leqslant t \leqslant x_i\}$. We note that φ, ψ are step functions, and so belong to L^1; further, $\varphi \geqslant g \geqslant \psi$ and

$$I(\varphi) = \sum_{i=1}^{n} M_i(x_i - x_{i-1}) = U(f, P),$$

$$I(\psi) = \sum_{i=1}^{n} m_i(x_i - x_{i-1}) = L(f, P).$$

(Remember that $\chi_{[x_{i-1}, x_i]}$, the characteristic function of the interval $[x_{i-1}, x_i]$, belongs to K^+ and $I(\chi_{[x_{i-1}, x_i]}) = x_i - x_{i-1}$. See §21.)

Let (P_n) be a sequence of partitions of $[a, b]$ such that P_{n+1} is a refinement of P_n ($n = 1, 2, \ldots$) and $\lim v(P_n) = 0$, $v(P_n)$ being the mesh of P_n. Let φ_n, ψ_n be the step functions associated with P_n. Since P_{n+1} is a refinement of P_n, we have

$$\varphi_n \geqslant \varphi_{n+1} \geqslant g \geqslant \psi_{n+1} \geqslant \psi_n.$$

Thus the sequences $(\varphi_n), (\psi_n)$ converge on $\mathbf{R}$, to F, G say. By the monotone convergence theorem, $F, G \in L^1$ and

$$I(F) = \lim I(\varphi_n), \quad I(G) = \lim I(\psi_n).$$

Since f is Riemann integrable over $[a, b]$ and $\lim v(P_n) = 0$, we have

$$\lim I(\varphi_n) = \lim U(f, P_n) = R\int_a^b f = \lim L(f, P_n) = \lim I(\psi_n).$$

Hence

$$I(F) = R\int_a^b f = I(G).$$

Since $I(F-G) = 0$ and $F \geqslant g \geqslant G$, 22.4 implies that $F = g = G$ a.e. Thus $g \in L^1$, i.e. f is Lebesgue integrable over $[a, b]$, and

$$L\int_a^b f = I(g) = I(F) = R\int_a^b f.$$

Above we defined Lebesgue integrability over a bounded interval $[a, b]$ and the Lebesgue integral over $[a, b]$. The corresponding definitions for an unbounded interval are given in a similar way.

We now discuss the relation between Lebesgue integrals and improper Riemann integrals. We consider only integrals over $\mathbf{R}$, but analogous results hold for other types of integral. We recall that a bounded real-valued function f on $\mathbf{R}$ is improperly Riemann integrable (over $\mathbf{R}$) if f is Riemann integrable over every bounded interval $[-s, t]$ and the limit $\lim_{s, t \to \infty} R\int_{-s}^{t} f$ exists and equals a real number. When this limit exists, we denote it by $R\int_{-\infty}^{\infty} f$.

23.2 Theorem. *Suppose f is Riemann integrable over every bounded interval and $|f|$ is improperly Riemann integrable. Then f is improperly Riemann integrable and Lebesgue integrable, and*

$$I(f) = R\int_{-\infty}^{\infty} f.$$

Proof. $(|f|\chi_{[-n,n]})$ is an increasing sequence of functions in L^1 which converges on **R** to $|f|$, and

$$I(|f|\chi_{[-n,n]}) = R\int_{-n}^{n}|f| \leq R\int_{-\infty}^{\infty}|f|.$$

Hence, by the monotone convergence theorem, $|f| \in L^1$.

Let (a_n), (b_n) be sequences of positive real numbers which diverge to ∞. Then $(f\chi_{[-a_n,b_n]})$ is a sequence of functions in L^1 which converges on **R** to f and

$$|f\chi_{[-a_n,b_n]}| \leq |f|.$$

Hence, by the dominated convergence theorem, $f \in L^1$ and

$$I(f) = \lim I(f\chi_{[-a_n,b_n]}) = \lim R\int_{-a_n}^{b_n}|f|.$$

Since this is true for any sequences (a_n), (b_n) of positive real numbers which diverge to ∞, the result follows.

23.3 THEOREM. *Suppose f belongs to L^1 and is Riemann integrable over every bounded interval. Then $|f|$ is improperly Riemann integrable.*

Proof. The hypotheses imply that $|f|$ belongs to L^1 and is Riemann integrable over every bounded interval. If (a_n), (b_n) are sequences of positive real numbers which diverge to ∞, then $(|f|\chi_{[-a_n,b_n]})$ is a sequence of functions in L^1 which converges on **R** to $|f|$ and is dominated by $|f|$. Hence

$$I(|f|) = \lim I(|f|\chi_{[-a_n,b_n]}) = \lim R\int_{-a_n}^{b_n}|f|.$$

The result follows.

We note that f may be improperly Riemann integrable without being Lebesgue integrable. For example, the improper Riemann integral $\int_{-\infty}^{\infty} \frac{\sin x}{x} dx$ is convergent but *not* absolutely convergent. Hence, by 23.3, the function $x \mapsto \frac{\sin x}{x}$ does not belong to L^1.

In view of the above results, we may drop the prefixes L and R from our integrals, provided we are careful not to use the convergence theorems of the Lebesgue theory when discussing conditionally convergent improper Riemann integrals.

We now look at some examples illustrating the monotone and dominated convergence theorems.

23.4 Examples. (a) If $\alpha > 1$, then

$$\int_0^\infty \frac{e^{-x}}{1-e^{-x}} x^{\alpha-1} \, dx = \Gamma(\alpha) \sum_{n=1}^\infty n^{-\alpha}$$

where $\Gamma(\alpha) = \int_0^\infty e^{-x} x^{\alpha-1} dx$.

If $x > 0$, $\dfrac{e^{-x}}{1-e^{-x}} = \sum_{k=1}^\infty e^{-kx}$. Write

$$s_n(x) = x^{\alpha-1} \sum_{k=1}^n e^{-kx} \quad (x \geq 0; \; n = 1, 2, \ldots).$$

Then (s_n) is an increasing sequence of functions Lebesgue integrable over $[0, \infty)$ and

$$\int_0^\infty s_n(x) \, dx = \sum_{k=1}^n \int_0^\infty e^{-kx} x^{\alpha-1} \, dx$$

$$= \sum_{k=1}^n k^{-\alpha} \int_0^\infty e^{-y} y^{\alpha-1} \, dy$$

$$= \Gamma(\alpha) \sum_{k=1}^n k^{-\alpha} < \Gamma(\alpha) \sum_{k=1}^\infty k^{-\alpha}.$$

Hence, by the monotone convergence theorem,

$$\int_0^\infty \frac{e^{-x}}{1-e^{-x}} x^{\alpha-1} dx = \int_0^\infty (\lim s_n(x)) \, dx$$

$$= \lim \int_0^\infty s_n(x) \, dx$$

$$= \Gamma(\alpha) \sum_{k=1}^\infty k^{-\alpha}.$$

(b) $\quad \lim \displaystyle\int_0^1 \dfrac{n^{3/2} x}{1+n^3 x^3} \, dx = 0.$

Write

$$f_n(x) = \frac{n^{3/2} x}{1+n^3 x^3} \quad (0 \leq x \leq 1; \; n = 1, 2, \ldots).$$

Then (f_n) is a sequence of functions Lebesgue integrable over $[0, 1]$ and $\lim f_n(x) = 0$ for each $x \in [0, 1]$. Hence the result will follow if either the monotone or the dominated convergence theorem is applicable. It is not hard to see that the sequence (f_n) is *not* monotonic. However, if x is a fixed point of $(0, 1]$,

$$\sup_{n \geq 1} |f_n(x)| \leq \sup_{t \geq 0} \frac{t^{3/2}x}{1+t^3 x^3} = \frac{1}{2\sqrt{x}},$$

as is easily shown using the differential calculus. Since the improper Riemann integral $\int_0^1 \frac{dx}{2\sqrt{x}}$ is (absolutely) convergent, the sequence (f_n) is dominated by a function Lebesgue integrable over $[0, 1]$. The conclusion follows.

(c) $\int_0^\infty e^{-xy} \frac{\sin y}{y} dy = \frac{1}{2}\pi - \arctan x$ for $x \geq 0$.

Write

$$\Phi(x) = \int_0^\infty e^{-xy} \frac{\sin y}{y} dy \quad (x \geq 0).$$

If $x > 0$, then $\left|\frac{\partial}{\partial x}\left(e^{-xy}\frac{\sin y}{y}\right)\right| = |-e^{-xy}\sin y| \leq e^{-xy}$ and the integral $\int_0^\infty e^{-xy} dy$ is convergent so that, by 22.8, we have,

$$\Phi'(x) = -\int_0^\infty e^{-xy} \sin y \, dy = \frac{-1}{1+x^2}$$

integrating by parts twice. Thus there is a number A such that

$$\Phi(x) = -\arctan x + A \quad (x > 0).$$

To find A, we note that

$$\left|e^{-xy}\frac{\sin y}{y}\right| \leq e^{-y} \quad \text{for } x \geq 1 \text{ and } y > 0.$$

Since the integral $\int_0^\infty e^{-y} dy$ is convergent, it follows from an obvious analogue of the dominated convergence theorem that

$$A - \tfrac{1}{2}\pi = \lim_{x \to \infty} \int_0^\infty e^{-xy} \frac{\sin y}{y} \, dy = \int_0^\infty \left(\lim_{x \to \infty} e^{-xy} \frac{\sin y}{y} \right) dy = 0.$$

Thus, if $x > 0$,

$$\Phi(x) = \tfrac{1}{2}\pi - \arctan x. \tag{23.2}$$

It must be emphasized that we have only established (23.2) for $x > 0$. We now show that $\Phi(0) = \tfrac{1}{2}\pi$, so that (23.2) is also valid when $x = 0$. It is sufficient to show that Φ is continuous at 0. (The argument

$$\tfrac{1}{2}\pi = \lim_{x \to 0+} \int_0^\infty e^{-xy} \frac{\sin y}{y} \, dy = \int_0^\infty \left(\lim_{x \to 0+} e^{-xy} \frac{\sin y}{y} \right) dy = \Phi(0)$$

cannot be justified using the dominated convergence theorem, for $\sup_{x > 0} \left| e^{-xy} \frac{\sin y}{y} \right| = \left| \frac{\sin y}{y} \right|$ and the integral $\int_0^\infty \left| \frac{\sin y}{y} \right| dy$ is divergent.)

We first note that if $x \geq 0$ and $t > 0$, then

$$\int_t^\infty e^{-xy} \frac{\sin y}{y} \, dy = e^{-xt} \frac{\cos t}{t} - \int_t^\infty \frac{(1+xy)}{y^2} e^{-xy} \cos y \, dy$$

(integrating by parts), so that

$$\left| \int_t^\infty e^{-xy} \frac{\sin y}{y} \, dy \right| \leq e^{-xt} \frac{|\cos t|}{t} + \int_t^\infty \frac{dy}{y^2} \leq \frac{2}{t}.$$

Let $\varepsilon > 0$. Then, as we have just seen, we can choose $t > 0$ such that

$$\left| \int_t^\infty e^{-xy} \frac{\sin y}{y} \, dy \right| < \tfrac{1}{3}\varepsilon \text{ for all } x \geq 0.$$

By the dominated convergence theorem,

$$\lim_{x \to 0+} \int_0^t e^{-xy} \frac{\sin y}{y} dy = \int_0^t \frac{\sin y}{y} dy.$$

Hence there is a $\delta > 0$ such that

$$\left| \int_0^t e^{-xy} \frac{\sin y}{y} dy - \int_0^t \frac{\sin y}{y} dy \right| < \tfrac{1}{3}\varepsilon$$

or all x satisfying $0 \leqslant x < \delta$. If x satisfies $0 \leqslant x < \delta$, then

$$|\Phi(x) - \Phi(0)| = \left| \int_0^\infty e^{-xy} \frac{\sin y}{y} dy - \int_0^\infty \frac{\sin y}{y} dy \right|$$

$$\leqslant \left| \int_0^t e^{-xy} \frac{\sin y}{y} dy - \int_0^t \frac{\sin y}{y} dy \right|$$

$$+ \left| \int_t^\infty e^{-xy} \frac{\sin y}{y} dy \right| + \left| \int_t^\infty \frac{\sin y}{y} dy \right|$$

$$< \tfrac{1}{3}\varepsilon + \tfrac{1}{3}\varepsilon + \tfrac{1}{3}\varepsilon = \varepsilon.$$

Thus Φ is continuous at 0, and $\Phi(x)$ is given by (23.2) for $x \geqslant 0$.

§24. Daniell integrals

Looking back on the material of §§20, 21 and 22, we see that our development of Lebesgue integration depends only on the properties of K and of the integral I on K mentioned in 20.3. This observation leads to the following abstraction.

24.1 DEFINITION. Let L be a collection of real-valued functions on a set X, and suppose that L is a real vector space and a lattice. (We shall refer to such a collection as a **vector lattice** of functions.) An increasing linear functional I on L, which is continuous under monotone limits, is called a **Daniell integral** on L.

24.2 *Examples.* (a) $X = \mathbf{R}$; $L_1 = K$, the collection of all continuous real-valued functions on **R** having compact support; I_1 defined by

$I_1(\varphi) =$ the Riemann integral of φ over any interval containing the support of φ.

(See 20.3.)

(b) $X = \mathbf{R}$; L_2, the collection of all step functions on **R**; I_2 defined by

$$I_2(\varphi) = \sum_{i=1}^{k} c_i(a_i - a_{i-1})$$

where $a_0 < a_1 < \ldots < a_k$ and

$$\varphi(x) = \begin{cases} 0 & \text{if } x < a_0 \text{ or } x > a_k \\ c_i & \text{if } a_{i-1} < x < a_i, \quad i = 1, \ldots, k: \end{cases}$$

$I_2(\varphi)$ is simply the Lebesgue integral of φ. (See Figure 24.1.)

It is easily seen that L_2 is a vector lattice of functions. That I_2 is a Daniell integral on L_2 follows from properties of the Lebesgue integral.

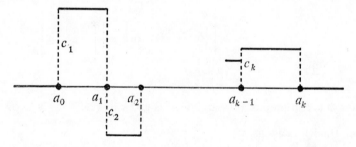

Figure 24.1. The graph of φ. The values of φ at the points a_i are irrelevant.

(c) $X = \mathbf{R}$; $L_3 = L^1$, the collection of all Lebesgue integrable functions on **R**; I_3 defined by

$I_3(\varphi) =$ the Lebesgue integral of φ.

(See 21.9 and 22.3.)

We shall introduce a further example in §27, when we discuss double integrals.

Let L be a vector lattice of functions on a set X, and I be a Daniell integral on L. Then, as we did in §§20 and 21, we can extend I to a

Daniell integral, which we again denote by I, on a larger collection of functions. We need only sketch the ideas. We define a subset E of X to be **null** (or more precisely I-**null**) if there is an increasing sequence (φ_n) of non-negative functions in L such that $\lim \varphi_n(x) = \infty$ for each $x \in E$ but $\sup I(\varphi_n) < \infty$. If a property holds for all points of X outside some I-null set, we say that the property holds a.e.(I). We denote by L^+ the collection of all real-valued functions f on X for which there is an increasing sequence (φ_n) of functions in L such that

$$\sup I(\varphi_n) < \infty \quad \text{and} \quad f = \lim \varphi_n \text{ a.e.}(I). \tag{24.1}$$

If $f \in L^+$, then $I(f)$ is well-defined by the equation

$$I(f) = \lim I(\varphi_n)$$

where (φ_n) is any increasing sequence of functions in L satisfying (24.1). We denote by L^1 the collection of all functions on X of the form $f_1 - f_2$ with $f_1, f_2 \in L^+$. The functions in L^1 are said to be I-**integrable**. If $f \in L^1$, say $f = f_1 - f_2$ where $f_1, f_2 \in L^+$, then $I(f)$ is well-defined by the equation

$$I(f) = I(f_1) - I(f_2). \tag{24.2}$$

L^1 is a vector lattice of functions, and I, defined by (24.2), is a Daniell integral on L^1. Analogues of the monotone and dominated convergence theorems hold. For example, the analogue of the monotone convergence theorem states: *if (f_n) is a monotonic sequence of functions in L^1 such that the sequence $(I(f_n))$ is bounded, then the sequence (f_n) converges a.e.(I) to a function f in L^1 and $I(f) = \lim I(f_n)$.*

This result shows that the Daniell integral I on L^1 cannot be extended, by our procedure, to a Daniell integral on a collection of functions properly containing L^1. Thus, in a sense, the collection of I-integrable functions is the natural domain of I.

An obvious question, arising from the above examples, is: What is the collection of integrable functions corresponding to the Daniell integral I_2? To answer this question, we prove the following general comparison theorem.

24.3 THEOREM. *Let L_1, L_2 be vector lattices of functions on the same set X, I_1, I_2 be Daniell integrals on L_1, L_2 respectively, and L_ν^1 be the set of all I_ν-integrable functions on X ($\nu = 1, 2$). Suppose that*

$$L_1 \subseteq L_2^1 \quad \text{and} \quad I_1(\varphi) = I_2(\varphi) \text{ for all } \varphi \in L_1.$$

Then

$$L_1^1 \subseteq L_2^1 \text{ and } I_1(f) = I_2(f) \text{ for all } f \in L_1^1.$$

Proof. If (φ_n) is an increasing sequence of functions in L_1 such that $\sup I_2(\varphi_n) = \sup I_1(\varphi_n) < \infty$, then by the monotone convergence theorem, $\{x \in X: \lim \varphi_n(x) = \infty\}$ is an I_2-null subset of X. It follows that every I_1-null subset of X is also I_2-null.

Now let $f \in L_1^1$, say $f = g - h$ where $g, h \in L_1^+$. Then there are increasing sequences (φ_n), (ψ_n) of functions in $L_1 \subseteq L_2^1$ which converge a.e.(I_1), and therefore a.e.(I_2), to g, h respectively. Since $I_2(\varphi_n) = I_1(\varphi_n) \leq I_1(g)$, the monotone convergence theorem implies that $g \in L_2^1$ and

$$I_2(g) = \lim I_2(\varphi_n) = \lim I_1(\varphi_n) = I_1(g).$$

Similarly $h \in L_2^1$ and $I_2(h) = I_1(h)$. Hence $f = g - h \in L_2^1$ and

$$I_2(f) = I_2(g) - I_2(h) = I_1(g) - I_1(h) = I_1(f).$$

24.4 COROLLARY. *If, in addition to the hypotheses of* 24.3, $L_2 \subseteq L_1^1$ *and* $I_1(\varphi) = I_2(\varphi)$ *for all* $\varphi \in L_2$, *then*

$$L_1^1 = L_2^1 \text{ and } I_1(f) = I_2(f) \text{ for all } f \in L_1^1 = L_2^1.$$

We can now answer the question raised earlier. Let I_1, I_2 be the Daniell integrals defined in 24.2(a), (b). We have already seen that every step function on **R** is Lebesgue integrable, i.e. $L_2 \subseteq L_1^1$, and $I_2(\varphi) = I_1(\varphi)$ for all $\varphi \in L_2$. We shall show that $L_1 \subseteq L_2^1$ and $I_1(\varphi) = I_2(\varphi)$ for all $\varphi \in L_1$.

Let $\varphi \in L_1 = K$, and $[a, b]$ be an interval containing the support of φ. With each partition $P: a = x_0 < x_1 < \ldots < x_n = b$ of $[a, b]$, we associate the function ψ on **R** defined by

$$\psi(x) = 0 \text{ if } x \in \mathbf{R} - [a, b], \ \psi(b) = \varphi(b),$$

$$\psi(x) = m_i \text{ if } x_{i-1} \leq x < x_i \quad (i = 1, 2, \ldots, n),$$

where $m_i = \inf \{\varphi(t): x_{i-1} \leq t \leq x_i\}$. ψ is a step function, $\psi \leq \varphi$ and $I_2(\psi) = L(\varphi, P) \leq I_1(\varphi)$. (See the proof of 23.1.) Let (P_n) be a sequence of partitions of $[a, b]$ such that P_{n+1} is a refinement of P_n $(n = 1, 2, \ldots)$, and $\lim v(P_n) = 0$. Let ψ_n be the step function associated with P_n. Then the sequence (ψ_n) is increasing, and, as we now show,

$$\lim \psi_n(x) = \varphi(x) \text{ for all } x \in \mathbf{R}, \qquad (24.3)$$

Both sides of (24.3) are 0 if $x \in \mathbf{R} - [a, b]$, so suppose $x \in [a, b]$, and let $\varepsilon > 0$. Since φ is continuous at x, there is a $\delta > 0$ such that $\varphi(y) > \varphi(x) - \varepsilon$ whenever $|x - y| < \delta$. Since $\lim v(P_n) = 0$, there is a positive integer n_0 such that $v(P_n) < \delta$ for all $n \geq n_0$. Then, if $n \geq n_0$,

$$\varphi(x) \geq \psi_n(x) \geq \varphi(x) - \varepsilon.$$

(24.3) follows. Hence $\varphi \in L_2^+ \subseteq L_2^1$ and

$$I_2(\varphi) = \lim I_2(\psi_n) = \lim L(\varphi, P_n) = I_1(\varphi).$$

24.4 now implies that

$$L_2^1 = L_1^1 \text{ and } I_2(f) = I_1(f) \text{ for all } f \in L_2^1.$$

Thus, in developing the theory of Lebesgue integration, we could have taken the Daniell integral I_2 of 24.2(b) as our starting point. This approach is adopted by several authors. It has the advantage that no knowledge of Riemann integration is assumed. The direct proof that I_2 is a Daniell integral is, however, rather irksome.

§25. Measurable functions and sets

In this section, we return to the discussion of the Lebesgue theory, though we bear in mind the possibility of generalizing our definitions and theorems to the Daniell setting of §24.

For some purposes the collection L^1 of Lebesgue integrable functions on $\mathbf{R}$ is not large enough. For example, L^1 is not closed under the operation of taking limits: the constant function 1 does not belong to L^1 despite being the limit of the sequence $(\chi_{[-n, n]})$ of Lebesgue integrable functions. We are led therefore to introduce a larger collection of functions, the Lebesgue measurable functions.

25.1 DEFINITION. A real-valued function f on $\mathbf{R}$ is said to be (**Lebesgue**) **measurable** if there is a sequence (φ_n) functions in K which converges a.e. to f.

The following facts about the collection of measurable functions are simple consequences of the definition.

25.2 THEOREM. (a) *If f is measurable and g is a real-valued function on $\mathbf{R}$ such that $g = f$ a.e., then g is measurable;*
(b) *every integrable function is measurable;*
(c) *the collection of measurable functions is a vector lattice;*

(d) *if f is measurable and there is an integrable function g such that $|f| \leq g$, then f is integrable;*
(e) *if f is measurable and c is a real number, then $f \wedge c$ is measurable.*

Proof. (a), (b) are obvious. (c), (e) follow since K is a vector lattice and, if $\varphi \in K$ and $c \in \mathbf{R}$, then $\varphi \wedge c \in K$. (d) is contained in 22.7.

We shall show presently that the collection of measurable functions is closed under the operation of taking limits. We first establish two lemmas, which are special cases of this result. The proofs of these lemmas are analogous to those of the monotone convergence theorem and the lemma preceeding it.

25.3 LEMMA. *Let (f_n) be an increasing sequence of functions in K^+ which converges a.e. to a function f. Then f is measurable.*

Proof. For each positive integer n, let $\varphi_{n1}, \varphi_{n2}, \ldots$ be an increasing sequence of functions in K which converges a.e. to f_n, and write

$$\psi_k = \varphi_{1k} \vee \varphi_{2k} \vee \ldots \vee \varphi_{kk} \quad (k = 1, 2, \ldots).$$

Then (ψ_k) is an increasing sequence of functions in K, and, for each positive integer k,

$$\varphi_{nk} \leq \psi_k \leq f_k \text{ a.e.*} \quad (n \leq k).$$

It follows that there is a function g such that $\lim \psi_k = g$ a.e. and

$$f_n \leq g \leq f \text{ a.e.*} \quad (n = 1, 2, \ldots). \tag{25.1}$$

By definition, g is measurable. Since $f = \lim f_n$ a.e., (25.1) implies that $f = g$ a.e.*, and hence f is measurable.

(The assertions marked * depend on the fact that the union of a countable collection of null sets is again a null set. See the proof of 22.1.)

25.4 LEMMA. *Let (f_n) be an increasing sequence of integrable functions which converges a.e. to a function f. Then f is measurable.*

Proof. For each positive integer n,

$$f_n = \sum_{i=1}^{n-1} (f_{i+1} - f_i) + f_1.$$

We can write $f_{i+1} - f_i = g_i - h_i$ where g_i, h_i are non-negative functions in K^+ and $I(h_i) < 2^{-i}$ $(i = 1, 2, \ldots)$. Then

$$f_n = G_n - H_n + f_1$$

where

$$G_n = \sum_{i=1}^{n-1} g_i \quad \text{and} \quad H_n = \sum_{i=1}^{n-1} h_i.$$

Now (G_n), (H_n) are increasing sequences of functions in K^+ and

$$I(H_n) = \sum_{i=1}^{n-1} I(h_i) < \sum_{i=1}^{n-1} 2^{-i} < 1.$$

Hence there is a function H in L^1 (in fact, by 22.1, $H \in K^+$) such that $H = \lim H_n$ a.e. Since $G_n = f_n + H_n - f_1$, the sequence (G_n) converges a.e. to a function G satisfying $G = f + H - f_1$ a.e. By 25.3, G is measurable. Hence f, which coincides a.e. with $G - H + f_1$, is measurable.

25.5 THEOREM. *Let (f_n) be a sequence of measurable functions which converges a.e. to a function f. Then f is measurable.*

Proof. Suppose first that f is non-negative. Then we may assume that each of the functions f_n is non-negative, otherwise we would replace f_n by $f_n \vee 0$. Let g be a strictly positive integrable function: for example we could take g to be the function defined by $g(x) = \exp(-|x|)$ $(x \in \mathbf{R})$. Write

$$g_n = f \wedge (ng) \quad (n = 1, 2, \ldots).$$

For each positive integer n, $(f_m \wedge (ng))_m$ is a sequence of integrable functions which converges a.e. to $f \wedge (ng)$ and is dominated by ng. Hence, by the dominated convergence theorem, $g_n \in L^1$ $(n = 1, 2, \ldots)$. The sequence (g_n) is increasing and converges to f. 25.4 therefore implies that f is measurable.

We now drop the restriction that f is non-negative. The above argument shows that the functions

$$f^+ = f \vee 0 = \lim (f_n \vee 0) \text{ a.e.}, f^- = (-f) \vee 0 = \lim ((-f_n) \vee 0) \text{ a.e.},$$

called the **positive** and **negative parts** of f, are measurable. Since $f = f^+ - f^-$, it follows that f is measurable.

25.6 COROLLARY. *A real-valued function f on $\mathbf{R}$ is measurable if and only if there is a sequence (f_n) of integrable functions which converges a.e. to f.*

We turn now to a discussion of measurable sets. If A is a subset of $\mathbf{R}$, we recall that the characteristic function of A, χ_A, is defined by

$$\chi_A(x) = \begin{cases} 1 & \text{if } x \in A \\ 0 & \text{if } x \in \mathbf{R} - A. \end{cases}$$

25.7 Definition. A subset A of $\mathbf{R}$ is said to be **(Lebesgue) measurable** if χ_A is a measurable function. Denote by $\mathcal{M}$ the collection of measurable subsets of $\mathbf{R}$. If $A \in \mathcal{M}$, write

$$\mu(A) = \begin{cases} I(\chi_A) & \text{if } \chi_A \text{ is integrable,} \\ \infty & \text{otherwise:} \end{cases}$$

$\mu(A)$ is called the **(Lebesgue) measure** of A.

The next result describes the main properties of $\mathcal{M}$ and μ.

25.8 Theorem. (a) *If (A_n) is a sequence of sets in $\mathcal{M}$, then $A_1 - A_2$ and $\bigcup_{n=1}^{\infty} A_n$ belong to $\mathcal{M}$;*
(b) *if $A \in \mathcal{M}$, then $\mu(A) \geq 0$ with equality if and only if A is a null set;*
(c) *if (A_n) is a sequence of pairwise disjoint sets in $\mathcal{M}$, then*

$$\mu\left(\bigcup_{n=1}^{\infty} A_n\right) = \sum_{n=1}^{\infty} \mu(A_n). \tag{25.2}$$

Proof. If $A \in \mathcal{M}$ and the function χ_A is integrable, then, since χ_A is non-negative, $\mu(A) = I(\chi_A) \geq 0$ and, by 22.4, $\mu(A) = 0$ if and only if $\chi_A = 0$ a.e., that is if and only if A is a null set. This establishes (b).

To prove (a) and (c), let (A_n) be a sequence of sets in $\mathcal{M}$, and write $A = \bigcup_{n=1}^{\infty} A_n$. Then

$$\chi_{A_1 - A_2} = \chi_{A_1} - \chi_{A_1 \cap A_2} = \chi_{A_1} - (\chi_{A_1} \wedge \chi_{A_2})$$

and

$$\chi_A = \lim (\chi_{A_1} \vee \chi_{A_2} \vee \ldots \vee \chi_{A_n})$$

are measurable functions, so that $A_1 - A_2, A \in \mathcal{M}$.

Suppose now that the sets $A_n, n = 1, 2, \ldots$, are pairwise disjoint. (25.2) is certainly true if $\mu(A_n) = \infty$ for some n. So suppose that each of the functions χ_{A_n} is integrable. Then $\left(\sum_{k=1}^{n} \chi_{A_k}\right)$ is an increasing sequence of integrable functions and

$$\chi_A = \lim\left(\sum_{k=1}^{n} \chi_{A_k}\right)$$

Hence, by the monotone convergence theorem, either χ_A is integrable and

$$\mu(A) = I(\chi_A) = \lim I\left(\sum_{k=1}^{n} \chi_{A_k}\right) = \sum_{k=1}^{\infty} \mu(A_k),$$

or χ_A is not integrable and

$$\mu(A) = \infty = \sum_{k=1}^{\infty} \mu(A_k).$$

25.9 Definition. A non-empty collection $\mathscr{S}$ of sets is called a σ-**ring** if $A_1 - A_2$ and $\bigcup_{n=1}^{\infty} A_n$ belong to $\mathscr{S}$ for any sequence (A_n) of sets in $\mathscr{S}$.

A function v on a σ-ring $\mathscr{S}$ of sets is called a **measure** if $0 \leqslant v(A) \leqslant \infty$ for each A in $\mathscr{S}$, $v(\varnothing) = 0$ and $v(\bigcup_{n=1}^{\infty} A_n) = \sum_{n=1}^{\infty} v(A_n)$ for any sequence (A_n) of pairwise disjoint sets in $\mathscr{S}$. v is said to be **complete** if every subset of a set of measure 0 belongs to $\mathscr{S}$.

According to 25.8, $\mathscr{M}$ is a σ-ring of subsets of **R** and μ is a measure. Since each subset of a null set is null and null sets are measurable, μ is complete.

We note that not every subset of **R** is measurable, though the construction of a non-measurable subset of **R** is non-trivial. (See Problem 4.9.) Thus the condition of measurability on a subset of **R** is a non-empty restriction.

The collection $\mathscr{M}$ of measurable subsets of **R** and the function μ on $\mathscr{M}$ were defined in terms of the measurable functions on **R** and the Lebesgue integral. It is interesting to see that conversely the measurable functions on **R** and the Lebesgue integral are determined by $\mathscr{M}$ and μ.

25.10 Theorem. *A real-valued function f on* **R** *is measurable if and only if the set* $\{x \in \mathbf{R} : f(x) > c\}$ *is measurable for each* $c \in \mathbf{R}$.

Proof. Suppose first that f is measurable, and let $c \in \mathbf{R}$. For each positive integer n, write

$$g_n = n[(f \wedge (c+n^{-1})) - (f \wedge c)].$$

Each of the functions g_n is measurable. If $f(x) \leqslant c$, then $g_n(x) = 0$; if $f(x) > c$, then, for all large n, $c+n^{-1} < f(x)$ and $g_n(x) = 1$. Thus the characteristic function of the set $\{x \in \mathbf{R}: f(x) > c\}$ is the limit of the sequence (g_n), and is therefore measurable.

Conversely, suppose that $\{x \in \mathbf{R}: f(x) > c\}$ is measurable for each $c \in \mathbf{R}$. Then, if $a, b \in \mathbf{R}$ and $a < b$,

$$\{x \in \mathbf{R}: b \geqslant f(x) > a\} = \{x \in \mathbf{R}: f(x) > a\} - \{x \in \mathbf{R}: f(x) > b\}$$

is measurable. For each positive integer n, let g_n be the function on $\mathbf{R}$ defined by

$$g_n(x) = \begin{cases} \dfrac{m}{n} & \text{if } \dfrac{m+1}{n} \geqslant f(x) > \dfrac{m}{n} \ (m = 0, \pm 1, \ldots, \pm n^2) \\ 0 & \text{otherwise.} \end{cases}$$

Since g_n is a linear combination of characteristic functions of measurable sets, it is measurable. If $x \in \mathbf{R}$ and $f(x) \in \left(-n, \dfrac{n^2+1}{n}\right]$, then $0 < f(x) - g_n(x) \leqslant n^{-1}$. Thus the sequence (g_n) converges to f, and f is measurable.

25.11 COROLLARY. (a) *If f is a measurable function and p is a positive real number, then $|f|^p$ is measurable;*
(b) *if f, g are measurable functions, then fg is measurable.*

Proof. (a) $\{x \in \mathbf{R}: |f(x)|^p > c\}$ equals $\mathbf{R}$ if $c < 0$ and equals $\{x \in \mathbf{R}: |f(x)| > \sqrt[p]{c}\}$ if $c \geqslant 0$.
(b) $fg = \frac{1}{4}[(f+g)^2 - (f-g)^2]$.

25.12 THEOREM. *Let f be a non-negative measurable function on $\mathbf{R}$. Then f is integrable if and only if*

$$\sup_{\varepsilon > 0} \sum_{n=1}^{\infty} n\varepsilon\mu(\{x \in \mathbf{R}: (n+1)\varepsilon \geqslant f(x) > n\varepsilon\}) = M < \infty, \quad (25.3)$$

and, in this case,

$$I(f) = M. \tag{25.4}$$

Proof. For each $\varepsilon > 0$ and each positive integer n, write

$$E_{\varepsilon n} = \{x \in \mathbf{R}: (n+1)\varepsilon \geqslant f(x) > n\varepsilon\}$$

and

$$f_\varepsilon = \sum_{n=1}^{\infty} n\varepsilon\chi_{\varepsilon n} = \lim_{p \to \infty}\left(\sum_{n=1}^{p} n\varepsilon\chi_{\varepsilon n}\right),$$

where $\chi_{\varepsilon n}$ denotes the characteristic function of $E_{\varepsilon n}$. Then f_ε is a non-negative measurable function and $f - \varepsilon \leqslant f_\varepsilon < f$, so that $f = \lim_{\varepsilon \to 0+} f_\varepsilon$.

Suppose that f is integrable. Then, since $0 \leqslant f_\varepsilon < f$, the dominated convergence theorem implies that f_ε is integrable, and

$$\sup_{\varepsilon > 0} I(f_\varepsilon) \leqslant I(f) = \lim_{\varepsilon \to 0+} I(f_\varepsilon) \leqslant \sup_{\varepsilon > 0} I(f_\varepsilon).$$

By the monotone convergence theorem,

$$I(f_\varepsilon) = \sum_{n=1}^{\infty} n\varepsilon I(\chi_{\varepsilon n}) = \sum_{n=1}^{\infty} n\varepsilon \mu(E_{\varepsilon n}),$$

and (25.3), (25.4) follow.

Conversely, suppose that (25.3) is satisfied. Then, by the monotone convergence theorem, f_ε is integrable for each $\varepsilon > 0$, and

$$I(f_\varepsilon) = \sum_{n=1}^{\infty} n\varepsilon \mu(E_{\varepsilon n}) \leqslant M.$$

Since $f = \lim_{\varepsilon \to 0+} f_\varepsilon$, Fatou's lemma, 22.5, implies that f is integrable.

At the beginning of this chapter, we mentioned that we would not follow Lebesgue's approach to integration theory. We are now in a position to give a brief outline of this approach. Lebesgue first showed that the concept of length, defined on the open subsets of **R**, can be extended to a complete measure μ on a σ-ring $\mathscr{M}$ of subsets of **R**. (The details of his argument are rather technical, and need not concern us. A uniqueness result ensures that μ and $\mathscr{M}$ are as we defined them in 25.7). He then defined a real-valued function f on **R** to be measurable if $\{x \in \mathbf{R}: f(x) > c\} \in \mathscr{M}$ for each $c \in \mathbf{R}$, a non-negative measurable function f on **R** to be integrable if

$$\sup_{\varepsilon > 0} \sum_{n=1}^{\infty} n\varepsilon \mu(\{x \in \mathbf{R}: (n+1)\varepsilon \geqslant f(x) > n\varepsilon\}) < \infty,$$

and, in this case, $I(f)$ to be this supremum. Finally, he defined a real-valued function f on **R** to be integrable if the non-negative functions f^+ and f^- are integrable according to the above definition, and, in this case, $I(f)$ to be $I(f^+) - I(f^-)$.

25.10 and 25.12 show that the integration theory so obtained is the same as that which we have presented. Our approach to Lebesgue integration, due to Daniell, has the advantage of directness—no initial discussion of measures is necessary.

We conclude this section by discussing briefly the extension of our results to the Daniell setting of §24. If I is a Daniell integral on a vector lattice L of functions on a set X, then a real-valued function f on X is said to be measurable (or more precisely I-measurable) if there is a sequence (φ_n) of functions in L which converges a.e. (I) to f. The elementary results about Lebesgue measurable functions, contained in 25.2 (a)-(d), clearly generalize to statements about I-measurable functions. In the proofs of 25.2(e) and 25.5, however, we used the following special properties of K and L^1:
(1) if $\varphi \in K$ and $c \in \mathbf{R}$, then $\varphi \wedge c \in K$;
(2) there is a function g in L^1 which is positive at each point of $\mathbf{R}$.
Actually, in the proof of 25.5, we only needed to know that if f is a non-negative function which is the limit a.e. of a sequence of measurable functions, then there is a non-negative integrable function g which is positive at each point at which f is positive. It is not difficult to show that this condition is satisfied in the general Daniell setting. If L satisfies
(1)' if $\varphi \in L$ and $c \in \mathbf{R}$, then $\varphi \wedge c \in L$,
known as **Stone's axiom**, then all of the results of this section have valid generalizations. We shall make use of this observation in §27.

§26. Complex-valued functions: L^p spaces
It is convenient, at this stage, to extend the concepts of integrability and measurability to complex-valued functions.

26.1 DEFINITION. A complex-valued function f is said to be **integrable (measurable)** if Re f and Im f are both integrable (measurable). Again denote by L^1 the collection of complex-valued integrable functions, and define I on L^1 by

$$I(f) = I(\mathrm{Re}\,f) + iI(\mathrm{Im}\,f) \qquad (f \in L^1).$$

26.2 THEOREM. (a) L^1 *is a complex vector space and I is a linear functional on L^1;*
(b) *if f is a complex-valued measurable function, then $f \in L^1$ if and only if $|f| \in L^1$;*
(c) *if $f \in L^1$, then $|I(f)| \leq I(|f|)$.*

Proof. (a) is easily verified.

To prove (b), let f be a complex-valued measurable function. If $f \in L^1$, then $\operatorname{Re} f, \operatorname{Im} f$ are real-valued functions in L^1. Since the real-valued functions in L^1 form a vector lattice, it follows that $|\operatorname{Re} f|, |\operatorname{Im} f| \in L^1$. Since $|f| = [(\operatorname{Re} f)^2 + (\operatorname{Im} f)^2]^{\frac{1}{2}}$ is measurable, the inequality $|f| \leq |\operatorname{Re} f| + |\operatorname{Im} f|$ implies that $|f| \in L^1$. Conversely if $|f| \in L^1$, then, since the functions $\operatorname{Re} f$ and $\operatorname{Im} f$ are both measurable, the inequalities $|\operatorname{Re} f|, |\operatorname{Im} f| \leq |f|$ show that $\operatorname{Re} f, \operatorname{Im} f \in L^1$ and so $f \in L^1$.

Finally to prove (c), let $f \in L^1$ and choose $\theta \in \mathbf{R}$ such that $e^{i\theta}I(f) \geq 0$. Then

$$\begin{aligned}
|I(f)| &= |e^{i\theta}I(f)| \\
&= e^{i\theta}I(f) \\
&= I(e^{i\theta}f) \\
&= I(\operatorname{Re}(e^{i\theta}f)) \quad (\operatorname{Im} I(e^{i\theta}f) = 0) \\
&\leq I(|f|) \quad (\operatorname{Re}(e^{i\theta}f) \leq |f| \text{ and } I \text{ is increasing}).
\end{aligned}$$

In many branches of mathematics it is useful to consider certain spaces of functions, the L^p spaces. We now define these spaces, and establish some of their important properties.

26.3 DEFINITION. For each real number $p \geq 1$, denote by $L^p = L^p(\mathbf{R})$ the collection of measurable functions f on $\mathbf{R}$ such that $|f|^p \in L^1$, and define $\|\cdot\|_p$ on L^p by

$$\|f\|_p = \{I(|f|^p)\}^{1/p} \quad (f \in L^p).$$

Since every function in L^1 is measurable, 26.2(b) shows that this definition of L^1 is consistent with the one given above.

We note that if $p > 1$, every bounded function in L^1 belongs to L^p. For if $f \in L^1$ and there is a number M such that $|f| \leq M$, then $|f|^p = |f|^{p-1}|f| \leq M^{p-1}|f|$. Thus there are lots of functions in L^p.

We shall show presently that L^p is a vector space and that $\|\cdot\|_p$ is almost a norm on L^p. We require:

26.4 LEMMA. *Let* $f \in L^p$ *and* $g \in L^q$ *where* $p > 1$ *and* $\dfrac{1}{p} + \dfrac{1}{q} = 1$. *Then* $fg \in L^1$ *and*

$$I(|fg|) \leq \|f\|_p \|g\|_q. \tag{26.1}$$

(*The inequality* (26.1) *is known as* **Hölder's inequality**.)

E

Proof. If $x \in \mathbf{R}$, then either

$$|g(x)| \leq |f(x)|^{p-1} \quad \text{or} \quad |f(x)| < |g(x)|^{1/(p-1)} = |g(x)|^{q-1}$$

Hence $|fg| \leq |f|^p \vee |g|^q \in L^1$. Since fg is measurable, it follows that $fg \in L^1$.

To prove (26.1), we first note that if a, b, α, β are non-negative real numbers and $\alpha + \beta = 1$, then

$$a^\alpha b^\beta \leq \alpha a + \beta b. \tag{26.2}$$

The inequality (26.2) is certainly true if $ab = 0$ and, when $ab \neq 0$, is equivalent to the inequality

$$\alpha \log a + \beta \log b \leq \log(\alpha a + \beta b),$$

which is valid since the graph of the logarithmic function is concave downwards. (See Figure 26.1.)

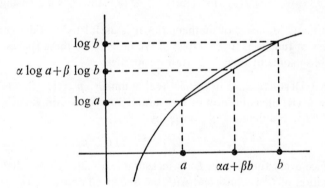

Figure 26.1

We now prove (26.1). If $\|f\|_p = 0$, then $f = 0$ a.e. and (26.1) is satisfied. Similarly (26.1) is satisfied if $\|g\|_q = 0$. Thus we may assume that $\|f\|_p \|g\|_q \neq 0$. If $x \in \mathbf{R}$, then, applying (26.2) with

$$a = \frac{|f(x)|^p}{\|f\|_p^p}, \quad b = \frac{|g(x)|^q}{\|g\|_q^q}, \quad \alpha = \frac{1}{p} \text{ and } \beta = \frac{1}{q},$$

we obtain

$$\frac{|f(x)g(x)|}{\|f\|_p \|g\|_q} \leq \frac{|f(x)|^p}{p \|f\|_p^p} + \frac{|g(x)|^q}{q \|g\|_q^q}.$$

Hence

$$\frac{I(|fg|)}{\|f\|_p \|g\|_q} \le \frac{1}{p}+\frac{1}{q} = 1,$$

and the result follows.

26.5 THEOREM. L^p *is a complex vector space. Further*
(a) *for each f in L^p and each complex number α*, $\|\alpha f\|_p = |\alpha| \|f\|_p$;
(b) *for all $f, g \in L^p$*,

$$\|f+g\|_p \le \|f\|_p + \|g\|_p. \qquad (26.3)$$

(*The inequality* (26.3) *is known as* **Minkowski's inequality**.)
Proof. If $f \in L^p$ and α is a complex number, then clearly $\alpha f \in L^p$ and (a) is satisfied.

Now suppose $f, g \in L^p$. Then, for each $x \in \mathbf{R}$,

$$\begin{aligned}|f(x)+g(x)|^p &\le \{|f(x)| + |g(x)|\}^p \\ &\le \{2 \max(|f(x)|, |g(x)|)\}^p \\ &= 2^p \max(|f(x)|^p, |g(x)|^p),\end{aligned}$$

so that $|f+g|^p \le 2^p(|f|^p \vee |g|^p) \in L^1$. It follows that $(f+g) \in L^p$.
If $p = 1$, Minkowski's inequality follows from the inequality $|f+g| \le |f| + |g|$. So suppose $p > 1$, and define q by $\frac{1}{p}+\frac{1}{q} = 1$.
Since $(p-1)q = p$, we have $|f+g|^{p-1} \in L^q$. Hence, by Hölder's inequality,

$$\begin{aligned}\|f+g\|_p^p &= I(|f+g|^p) \\ &\le I(|f| |f+g|^{p-1}) + I(|g| |f+g|^{p-1}) \\ &\le \|f\|_p \|f+g\|_p^{p/q} + \|g\|_p \|f+g\|_p^{p/q}. \qquad (26.4)\end{aligned}$$

If $\|f+g\|_p = 0$, the desired inequality is obvious. Otherwise we may divide through by $\|f+g\|_p^{p/q}$ in (26.4). Since $p - \frac{p}{q} = 1$, we obtain $\|f+g\|_p \le \|f\|_p + \|g\|_p$.

We note that $\|\cdot\|_p$ is *not* a norm on L^p for $\|f\|_p = 0$ does not imply that $f = 0$, but only $f = 0$ a.e. However, if we identify functions which differ only on a null set, then $(L^p, \|\cdot\|_p)$ becomes a normed space. More formally, the relation $\sim$ on L^p defined by

$$f \sim g \text{ if and only if } f = g \text{ a.e.}$$

is an equivalence relation. If $f \in L^p$, denote the equivalence class of f by $[f]$ and the set of equivalence classes into which L^p is divided by $\mathscr{L}^p$. If $f \sim f_1, g \sim g_1$ and α is a complex number, then $f+g \sim f_1+g_1$ and $\alpha f \sim \alpha f_1$. Hence we may define unambiguously

$$[f]+[g] = [f+g] \quad \text{and} \quad \alpha[f] = [\alpha f].$$

With these definitions, $\mathscr{L}^p$ becomes a vector space. $\mathscr{L}^p$ becomes a normed space if we define $\|[f]\|_p = \|f\|_p$: $\|[f]\|_p = 0$ implies $f = 0$ a.e., i.e. $[f] = [0]$, the zero of $\mathscr{L}^p$. It is customary not to insist on the distinction between L^p and $\mathscr{L}^p$.

26.6 THEOREM. *The space* $(L^p, \|\cdot\|_p)$ *is complete.*

Proof. Let (f_n) be a Cauchy sequence in this space. Then, for each positive integer n, there is a positive integer k_n such that

$$\|f_r - f_s\|_p < 2^{-n} \text{ for all } r, s \geq k_n.$$

Clearly we may assume that the sequence (k_n) is strictly increasing, so that (f_{k_n}) is a subsequence of (f_n). Now, for each positive integer N, we have, using Minkowski's inequality,

$$\left(I\left(\sum_{n=1}^N |f_{k_{n+1}} - f_{k_n}|\right)^p\right)^{1/p} \leq \sum_{n=1}^N \|f_{k_{n+1}} - f_{k_n}\|_p < \sum_{n=1}^N 2^{-n} < 1.$$

Hence, by the monotone convergence theorem, $\sum_{n=1}^\infty |f_{k_{n+1}} - f_{k_n}|$ converges a.e. It follows that $\sum_{n=1}^\infty (f_{k_{n+1}} - f_{k_n})$ converges a.e. Since $\sum_{n=1}^N (f_{k_{n+1}} - f_{k_n}) = f_{k_{N+1}} - f_{k_1}$, this implies that the sequence (f_{k_n}) converges a.e. Let f be any function on **R** which is the limit a.e. of this sequence. Then, if m is a fixed positive integer,

$$|f - f_{k_m}|^p = \lim_{n \to \infty} |f_{k_n} - f_{k_m}|^p \text{ a.e.}$$

and

$$I(|f_{k_n} - f_{k_m}|^p) \leq 2^{-mp} \text{ for all } n \geq m.$$

Hence, by Fatou's lemma, $|f - f_{k_m}|^p \in L^1$, i.e. $f - f_{k_m} \in L^p$, and $f = (f - f_{k_m}) + f_{k_m} \in L^p$. Further

$$\|f - f_{k_m}\|_p^p = I(|f - f_{k_m}|^p) \leq 2^{-mp}.$$

We now show that (f_n) converges to f with respect to $\|\cdot\|_p$. Let $\varepsilon > 0$, and choose a positive integer m such that $2^{-m} < \tfrac{1}{2}\varepsilon$ and $\|f_r - f_s\|_p < \tfrac{1}{2}\varepsilon$ for all $r, s \geq m$. Then, if $n \geq m$,

$$\begin{aligned}\|f - f_n\|_p &\leq \|f - f_{k_m}\|_p + \|f_{k_m} - f_n\|_p \\ &< 2^{-m} + \tfrac{1}{2}\varepsilon \quad \text{(since } k_m \geq m) \\ &< \varepsilon.\end{aligned}$$

The theorem follows.

The proof of 26.6 implies the following corollary, which we shall require later.

26.7 COROLLARY. *If (f_n) is a Cauchy sequence in L^p which converges in L^p to f, there is a strictly increasing sequence (k_n) of positive integers such that (f_{k_n}) converges to f a.e.*

26.8 THEOREM. *The set of continuous complex-valued functions on $\mathbf{R}$ with compact support is dense in L^p, i.e. if $f \in L^p$ and $\varepsilon > 0$, there is a continuous complex-valued function φ on $\mathbf{R}$ with compact support such that $\|f - \varphi\|_p < \varepsilon$.*

Proof. Let $f \in L^p$ and $\varepsilon > 0$. Since

$$f = (\operatorname{Re} f)^+ - (\operatorname{Re} f)^- + i[(\operatorname{Im} f)^+ - (\operatorname{Im} f)^-] \text{ and } (\operatorname{Re} f)^\pm, (\operatorname{Im} f)^\pm \in L^p,$$

it is sufficient to prove that K is dense in the subset of L^p consisting of non-negative functions. We first consider the case in which $p = 1$.

Suppose that f is a non-negative function in L^1. Then f has a decomposition $f = f_1 - f_2$ where $f_1, f_2 \in K^+$. There are increasing sequences $(\varphi_{1n}), (\varphi_{2n})$ of functions in K such that, for $i = 1, 2$,

$$f_i = \lim \varphi_{in} \text{ a.e.} \quad \text{and} \quad I(f_i) = \lim I(\varphi_{in}),$$

so that

$$\|f_i - \varphi_{in}\|_1 = I(|f_i - \varphi_{in}|) - I(f_i - \varphi_{in}) \to 0.$$

Write

$$\varphi_n = \varphi_{1n} - \varphi_{2n} \quad (n = 1, 2, \ldots).$$

Then (φ_n) is a sequence of functions in K and

$$\|f - \varphi_n\|_1 \leq \|f_1 - \varphi_{1n}\|_1 + \|f_2 - \varphi_{2n}\|_1 \to 0.$$

Hence there is a function φ in K such that $\|f - \varphi\|_1 < \varepsilon$. We may assume that φ is non-negative, otherwise we could replace φ by φ^+; φ^+ is a non-negative function in K and, as f is non-negative, $|f-\varphi^+| \leq |f-\varphi|$, so that $\|f-\varphi^+\|_1 \leq \|f-\varphi\|_1$. Further, if there is a real number c such that $f \leq c$, we may assume that $\varphi \leq c$ also, otherwise we would replace φ by $\varphi \wedge c$.

Now suppose that f is a non-negative function in L^p where $p > 1$. For each positive integer n, write

$$E_n = \{x \in \mathbf{R}: n^{-1} < f(x) \leq n\} \quad \text{and} \quad f_n = f\chi_{E_n}.$$

Then E_n is a measurable set, f_n is a measurable function and $0 \leq f_n \leq n$. If $\dfrac{1}{p}+\dfrac{1}{q} = 1$, then $\chi_{E_n}^q = \chi_{E_n} \leq n^p f^p$, so that $\chi_{E_n} \in L^q$ and $f_n \in L^1$. Since $E_n \subseteq E_{n+1}$, the sequence (f_n) is increasing. Clearly $f = \lim f_n$. Combining these facts, we see that $((f-f_n)^p)$ is a decreasing sequence of functions in L^1 which converges to the zero function, and so, by the monotone convergence theorem, $\lim I((f-f_n)^p) = 0$. Hence we may choose n such that $\|f-f_n\|_p < \tfrac{1}{2}\varepsilon$. According to the above result, there is a function φ in K such that

$$0 \leq \varphi \leq n \quad \text{and} \quad \|f_n-\varphi\|_1 < n^{1-p}(\tfrac{1}{2}\varepsilon)^p.$$

Then

$$\|f_n-\varphi\|_p^p = I(|f_n-\varphi|^p) = I(|f_n-\varphi|^{p-1}|f_n-\varphi|)$$
$$\leq n^{p-1}I(|f_n-\varphi|) < (\tfrac{1}{2}\varepsilon)^p,$$

so that $\|f_n-\varphi\|_p < \tfrac{1}{2}\varepsilon$. Finally

$$\|f-\varphi\|_p \leq \|f-f_n\|_p + \|f_n-\varphi\|_p < \tfrac{1}{2}\varepsilon+\tfrac{1}{2}\varepsilon = \varepsilon.$$

Denote by K' the set of continuous complex-valued functions on $\mathbf{R}$ with compact support. K' is a complex vector space and $\|\cdot\|_p$ is a genuine norm on K': if φ and ψ are distinct functions in K', then $\{x \in \mathbf{R}: \varphi(x) \neq \psi(x)\}$ is a non-empty open subset of $\mathbf{R}$ and $\|\varphi-\psi\|_p > 0$. It can be shown that this normed space is not complete. (See the final argument in §5.) According to 26.6 and 26.8, the space L^p is complete and K' is a dense subset of L^p. Thus L^p is, in a rather natural sense, the "completion" of K' under the norm $\|\cdot\|_p$. In particular, the completion of the normed space $(K', \|\cdot\|_1)$ is the space of Lebesgue integrable functions on $\mathbf{R}$ with any two such functions identified if they are equal a.e. This result is one of the main motivations of Lebesgue integration.

§27. Double integrals

In this section, we shall use the Daniell extension procedure, described in §24, to develop the theory of Lebesgue integration for functions on $\mathbf{R}^2$. In view of the way in which we defined the Lebesgue integrable functions on $\mathbf{R}$, it would be natural to take as the initial vector lattice of functions the continuous real-valued functions on $\mathbf{R}^2$ with compact support, and, for such a function Ω, to define $I(\Omega)$ to be the Riemann integral of Ω over any rectangle containing the support of Ω. However, as some readers may be unfamiliar with the theory of Riemann integration for functions on $\mathbf{R}^2$, we shall adopt a different starting point.

Denote by L the collection of functions Ω on $\mathbf{R}^2$ such that there are step functions $\varphi_1, \ldots, \varphi_k, \psi_1, \ldots, \psi_k$ on $\mathbf{R}$ with

$$\Omega(x, y) = \sum_{j=1}^{k} \varphi_j(x)\psi_j(y) \quad ((x, y) \in \mathbf{R}^2). \tag{27.1}$$

We note that L satisfies Stone's axiom, i.e. if $\Omega \in L$ and $c \in \mathbf{R}$, then $\Omega \wedge c \in L$. If $\Omega \in L$ and satisfies (27.1), where φ_j, ψ_j ($j = 1, \ldots, k$) are step functions on $\mathbf{R}$, then $\int_{-\infty}^{\infty} \Omega(x, \cdot) dx = \sum_{j=1}^{k} (\int_{-\infty}^{\infty} \varphi_j(x) dx) \psi_j$ is a step function on $\mathbf{R}$ and

$$\int_{-\infty}^{\infty} \left(\int_{-\infty}^{\infty} \Omega(x, y) dx \right) dy = \sum_{j=1}^{k} \left(\int_{-\infty}^{\infty} \varphi_j(x) dx \right) \left(\int_{-\infty}^{\infty} \psi_j(y) dy \right).$$

Similarly

$$\int_{-\infty}^{\infty} \left(\int_{-\infty}^{\infty} \Omega(x, y) dy \right) dx = \sum_{j=1}^{k} \left(\int_{-\infty}^{\infty} \varphi_j(x) dx \right) \left(\int_{-\infty}^{\infty} \psi_j(y) dy \right).$$

We define

$$I(\Omega) = \int_{-\infty}^{\infty} \left(\int_{-\infty}^{\infty} \Omega(x, y) dx \right) dy = \int_{-\infty}^{\infty} \left(\int_{-\infty}^{\infty} \Omega(x, y) dy \right) dx. \tag{27.2}$$

From now on, we shall, for convenience, drop the limits of integration.

27.1 THEOREM. *L is a vector lattice of functions and the function I, defined on L by (27.2), is a Daniell integral.*

Proof. Clearly L is a real vector space. To prove that L is a lattice, we need only show that if $\Omega \in L$, then $|\Omega| \in L$. So let $\Omega \in L$, say Ω is given by (27.1) where the functions φ_j, ψ_j are step functions on $\mathbf{R}$.

Then $\{(\psi_1(x), \psi_2(x), \ldots, \psi_k(x)): x \in \mathbf{R}\}$ is a finite subset of $\mathbf{R}^k$. Denote the points of this set, different from $(0, 0, \ldots, 0)$, by $(a_{1p}, a_{2p}, \ldots, a_{kp})$, $p = 1, \ldots, q$. Then, for $p = 1, \ldots, q$,

$$E_p = \{x \in \mathbf{R}: \psi_j(x) = a_{jp} \text{ for } j = 1, \ldots, k\}$$

is the union of a finite number of bounded intervals, and χ_p, the characteristic function of E_p, is a step function on $\mathbf{R}$. For $j = 1, \ldots, k$,

$$\psi_j = \sum_{p=1}^{q} a_{jp}\chi_p,$$

and so, for all $(x, y) \in \mathbf{R}^2$, we have

$$\Omega(x, y) = \sum_{j=1}^{k} \varphi_j(x) \left(\sum_{p=1}^{q} a_{jp}\chi_p(y) \right)$$
$$= \sum_{p=1}^{q} \left(\sum_{j=1}^{k} a_{jp}\varphi_j(x) \right) \chi_p(y) = \sum_{p=1}^{q} \theta_p(x)\chi_p(y)$$

where $\theta_p = \sum_{j=1}^{k} a_{jp}\varphi_j$ is a step function on $\mathbf{R}$. Since the sets $E_1, \ldots, E_p$ are disjoint,

$$|\Omega(x, y)| = \sum_{p=1}^{q} |\theta_p(x)| \chi_p(y) \quad ((x, y) \in \mathbf{R}^2).$$

It follows that $|\Omega| \in L$.

Since $I(\Omega) = \int (\int \Omega(x, y) dx) dy$ $(\Omega \in L)$, it is clear that I is an increasing linear functional on L. To show that I is continuous under monotone limits, let (Ω_n) be a decreasing sequence of functions in L which converges on $\mathbf{R}^2$ to the zero function. Then $(\int \Omega_n(x, \cdot) dx)$ is a decreasing sequence of step functions on $\mathbf{R}$ which, by the monotone convergence theorem, converges to the zero function. Hence, by the monotone convergence theorem again,

$$\lim I(\Omega_n) = \lim \int (\int \Omega_n(x, y) dx) dy = 0.$$

27.2 Definition. Denote by $L^1(\mathbf{R}^2)$ the collection of complex-valued integrable functions on $\mathbf{R}^2$ determined by the above Daniell integral. The functions in $L^1(\mathbf{R}^2)$ are said to be **(Lebesgue) integrable**.

If $\Omega \in L$, then

$$I(\Omega) = \int (\int \Omega(x, y) dx) dy = \int (\int \Omega(x, y) dy) dx.$$

We shall presently show that these equations are valid for all integrable functions, i.e. if $F \in L^1(\mathbf{R}^2)$, then the two repeated integrals $\int (\int F(x,y)dx)dy$, $\int (\int F(x,y)dy)dx$ exist and equal $I(F)$. We first define precisely what we mean by the existence of these repeated integrals.

27.3 DEFINITION. Let F be a complex-valued function on $\mathbf{R}^2$. Then the repeated integral $\int (\int F(x,y)dx)dy$ is said to exist if $F(\cdot, y)$ is a Lebesgue integrable function on $\mathbf{R}$ for a.a. y in $\mathbf{R}$ and there is a Lebesgue integrable function g on $\mathbf{R}$ such that

$$g(y) = \int F(x,y)dx \text{ for a.a. } y \text{ in } \mathbf{R}. \qquad (27.3)$$

The existence of the repeated integral $\int (\int F(x,y)dy)dx$ is defined in a similar way.

If the repeated integral $\int (\int F(x,y)dx)dy$ exists and g is a function on $\mathbf{R}$ satisfying (27.3), we shall denote the value of the integral $\int g(y)dy$ by $\int (\int F(x,y)dx)dy$, even though the function $\int F(x, \cdot)dx$ may only be defined a.e.

27.4 LEMMA. *Let E be a null subset of $\mathbf{R}^2$. Then, for a.a. y in $\mathbf{R}$, the section $E_y = \{x \in \mathbf{R}: (x,y) \in E\}$ is a null subset of $\mathbf{R}$.*

Proof. By definition, there is an increasing sequence (Ω_n) of nonnegative functions in L such that

$$\sup I(\Omega_n) < \infty \text{ and } \lim \Omega_n(x,y) = \infty \text{ for all } (x,y) \in E.$$

Now $(\int \Omega_n(x, \cdot)dx)$ is an increasing sequence of step functions on $\mathbf{R}$ and $\sup \int (\int \Omega_n(x,y)dx)dy = \sup I(\Omega_n) < \infty$. Hence, by the monotone convergence theorem,

$$S = \{y \in \mathbf{R}: \lim \int \Omega_n(x,y)dx = \infty\}$$

is a null set. If $y \notin S$, then $(\Omega_n(\cdot, y))$ is an increasing sequence of step functions, $\sup \int \Omega_n(x,y)dx < \infty$ and $\lim \Omega_n(x,y) = \infty$ for all $x \in E_y$, so that E_y is a null set.

27.5 FUBINI'S THEOREM. *Let $F \in L^1(\mathbf{R}^2)$. Then the two repeated integrals $\int (\int F(x,y)dx)dy$, $\int (\int F(x,y)dy)dx$ exist and equal $I(F)$.*

Proof. Since F has a decomposition $F = F_1 - F_2 + i(F_3 - F_4)$ where $F_1, F_2, F_3, F_4 \in L^+$, it is sufficient to prove the result when

$F \in L^+$. In this case, there is an increasing sequence (Ω_n) of functions in L such that
$$E = \{(x, y) \in \mathbf{R}^2 : \lim \Omega_n(x, y) \neq F(x, y)\}$$
is a null set and $\sup I(\Omega_n) < \infty$. Then, as in the proof of 27.4, we have
$$\sup \int \Omega_n(x, y)\, dx < \infty \text{ for a.a. } y \text{ in } \mathbf{R}.$$
Since, for each y in $\mathbf{R}$, $(\Omega_n(\cdot, y))$ is an increasing sequence of step functions on $\mathbf{R}$ and, by 27.4, for a.a. y in $\mathbf{R}$,
$$\{x \in \mathbf{R} : \lim \Omega_n(x, y) \neq F(x, y)\} = \{x \in \mathbf{R} : (x, y) \in E\}$$
is a null set, we see that, for a.a. y in $\mathbf{R}$, $F(\cdot, y)$ is integrable and
$$\int F(x, y)\, dx = \lim \int \Omega_n(x, y)\, dx.$$
Since $(\int \Omega_n(x, \cdot)\, dx)$ is an increasing sequence of step functions on $\mathbf{R}$ and $\sup \int (\int \Omega_n(x, y)\, dx)\, dy = \sup I(\Omega_n) < \infty$, there is an integrable function g on $\mathbf{R}$ such that, for a.a. y in $\mathbf{R}$,
$$g(y) = \lim \int \Omega_n(x, y)\, dx = \int F(x, y)\, dx.$$
Thus the repeated integral $\int (\int F(x, y)\, dx)\, dy$ exists. Further
$$\int g(y)\, dy = \lim \int (\int \Omega_n(x, y)\, dx)\, dy = \lim I(\Omega_n) = I(F).$$
The other half of the theorem is proved in a similar way.

Our next result is a very useful complement to Fubini's theorem.

27.6 Tonelli's Theorem. *Let F be a measurable function on $\mathbf{R}^2$, and suppose that one of the repeated integrals $\int (\int |F(x, y)|\, dx)\, dy$, $\int (\int |F(x, y)|\, dy)\, dx$ exists. Then F is integrable, and so the repeated integrals $\int (\int F(x, y)\, dx)\, dy$, $\int (\int F(x, y)\, dy)\, dx$ both exist and equal $I(F)$.*

Proof. Suppose that the repeated integral $\int (\int |F(x, y)|\, dx)\, dy$ exists. For each positive integer n, let χ_n be the characteristic function of the square $\{(x, y) \in \mathbf{R}^2 : -n \leqslant x, y \leqslant n\}$. Then $(|F| \wedge n\chi_n)$ is an increasing sequence of measurable functions which converges on $\mathbf{R}^2$ to $|F|$. Since $|F| \wedge n\chi_n \leqslant n\chi_n$, an integrable function, the functions $|F| \wedge n\chi_n$ are, in fact, integrable. By Fubini's theorem,
$$I(|F| \wedge n\chi_n) = \int (\int \min\{|F(x, y)|, n\chi_n(x, y)\}\, dx)\, dy$$
$$\leqslant \int (\int |F(x, y)|\, dx)\, dy.$$

LEBESGUE INTEGRATION

Hence, by the monotone convergence theorem, $|F| \in L^1(\mathbf{R}^2)$, which, since F is measurable, implies that $F \in L^1(\mathbf{R}^2)$.

Tonelli's theorem provides a simple, but very useful, way of justifying the change of order of integration in a repeated integral.

27.7 Examples. (a) Write $F(x, y) = y \exp(-(1+x^2)y^2)$. Then

$$\int_0^\infty F(x, y) \, dy = \left[\frac{-\exp(-(1+x^2)y^2)}{2(1+x^2)} \right]_0^\infty = \frac{1}{2(1+x^2)}$$

and

$$\int_0^\infty \frac{dx}{2(1+x^2)} = \tfrac{1}{2}[\arctan x]_0^\infty = \tfrac{1}{4}\pi.$$

Now F, being continuous, is measurable. (See Problem 4.20.) Hence, applying Tonelli's theorem to the non-negative measurable function $F\chi$, where χ is the characteristic function of the set $\{(x, y) \in \mathbf{R}^2 : x, y \geq 0\}$, we see that

$$\int_0^\infty \left(\int_0^\infty F(x, y) \, dx \right) dy = \int_0^\infty \left(\int_0^\infty F(x, y) \, dy \right) dx = \tfrac{1}{4}\pi.$$

But

$$\int_0^\infty \left(\int_0^\infty F(x, y) \, dx \right) dy = \int_0^\infty y \exp(-y^2) \left(\int_0^\infty \exp(-x^2 y^2) \, dx \right) dy$$

$$= \left(\int_0^\infty \exp(-y^2) \, dy \right) \left(\int_0^\infty \exp(-t^2) \, dt \right)$$

by an obvious change of variable. Thus

$$\int_0^\infty \exp(-y^2) \, dy = \tfrac{1}{2}\sqrt{\pi}.$$

(b) Let $f, g \in L^1(\mathbf{R})$. Then there is a function h in $L^1(\mathbf{R})$ such that, for a.a. x in $\mathbf{R}$,

$$h(x) = \int_{-\infty}^\infty f(x-y)g(y) \, dy;$$

further $\|h\|_1 \leq \|f\|_1 \|g\|_1$. (This element h of $L^1(\mathbf{R})$ is called the **convolution** of f and g, and denoted by $f * g$.)

We first prove that the function H on $\mathbf{R}^2$ defined by

$$H(x, y) = f(x-y)g(y) \quad ((x, y) \in \mathbf{R}^2)$$

is measurable. It is easily seen that the function $(x, y) \mapsto g(y)$ is measurable, and so the measurability of H will follow if we show that the function $(x, y) \mapsto f(x-y)$ is measurable.

If φ is a function on $\mathbf{R}$, define $T\varphi$ by

$$(T\varphi)(x, y) = \varphi(x-y) \quad ((x, y) \in \mathbf{R}^2).$$

Then $T\chi_{[a, b]}$ is the characteristic function of the strip S shown in Figure 27.1. Since the function χ_n defined on $\mathbf{R}^2$ by

$$\chi_n(x, y) = \sum_{j=-n^2}^{n^2-1} \chi_{j^1}(x)\, \chi_{j^2}(y),$$

where x_{j^1}, x_{j^2}, are the characteristic functions of the intervals $[a+j/n,\ b+(j+1)/n]$, $[j/n,\ (j+1)/n]$ respectively, belongs to $L^1(\mathbf{R}^2)$ and $\lim \chi_n = T\chi_{[a, b]}$, we see that $T\chi_{[a, b]}$ is measurable. If φ is a step function on $\mathbf{R}$, then it is a linear combination of characteristic functions of closed intervals, and so, since T is clearly linear, $T\varphi$ is measurable. Now since f is measurable, there is a sequence (φ_n) of step functions on $\mathbf{R}$ such that the set

$$E = \{x \in \mathbf{R} : f(x) \neq \lim \varphi_n(x)\}$$

is null. Then $(T\varphi_n)$ is a sequence of measurable functions on $\mathbf{R}^2$ and

$$(Tf)(x, y) = f(x-y) = \lim \varphi_n(x-y) = \lim (T\varphi_n)(x, y)$$

at all points (x, y) for which $(x-y) \notin E$. Since, by Problem 4.23,

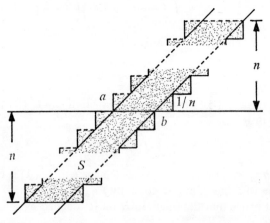

Figure 27.1. χ_n is the characteristic function of the shaded region.

$\{(x, y) \in \mathbf{R}^2 : x - y \in E\}$ is a null set, it follows that Tf is measurable and hence H is measurable.

Now

$$\int_{-\infty}^{\infty} \left(\int_{-\infty}^{\infty} |f(x-y)g(y)| \, dx \right) dy = \int_{-\infty}^{\infty} |g(y)| \left(\int_{-\infty}^{\infty} |f(x-y)| \, dx \right) dy$$

$$= \|f\|_1 \|g\|_1.$$

Hence, by Tonelli's theorem, the repeated integral

$$\int_{-\infty}^{\infty} \left(\int_{-\infty}^{\infty} f(x-y)g(y) \, dy \right) dx$$

exists, i.e. there is a function h in $L^1(\mathbf{R})$ such that, for a.a. x in $\mathbf{R}$,

$$h(x) = \int_{-\infty}^{\infty} f(x-y)g(y) \, dy.$$

Further

$$\|h\|_1 = \int_{-\infty}^{\infty} \left| \int_{-\infty}^{\infty} f(x-y)g(y) \, dy \right| dx$$

$$\leq \int_{-\infty}^{\infty} \left(\int_{-\infty}^{\infty} |f(x-y)g(y)| \, dy \right) dx$$

$$= \int_{-\infty}^{\infty} \left(\int_{-\infty}^{\infty} |f(x-y)g(y)| \, dx \right) dy = \|f\|_1 \|g\|_1.$$

We note that Problems 4.14 and 4.15 describe other situations in which the convolution $f * g$ of two functions f, g on $\mathbf{R}$ is defined by the equation

$$(f * g)(x) = \int_{-\infty}^{\infty} f(x-y)g(y) \, dy \quad (x \in \mathbf{R}).$$

Problems on Chapter 4

1. Let E be the set of all real numbers in $(0, 1)$ which do not have a decimal expansion containing the digit 5. Show that
(a) E is a null set, and
(b) E is not countable.
[Hint: to prove (a), note, as a first step, that

$$E \subseteq (0, 1) - [\cdot 5, \cdot 6] = (0, \cdot 5) \cup (\cdot 6, 1).]$$

2. If $f, g \in K^+$, show that $f \vee g \in K^+$.

3. Let $f \in L^1(\mathbf{R})$, $a \in \mathbf{R}$ and define the function f_a by
$$f_a(x) = f(x+a) \quad (x \in \mathbf{R}).$$
Show that $f_a \in L^1(\mathbf{R})$ and $I(f_a) = I(f)$.

4. If $f_n = -n^{-1} \chi_{[0, n]}$, show that the sequence (f_n) converges uniformly on $\mathbf{R}$ to the zero function and $\liminf \int_{-\infty}^{\infty} f_n(x)dx = -1$. Why does this not contradict Fatou's lemma?

5. Let (f_n) be the sequence of functions on $\mathbf{R}$ defined by
$$f_{2n-1}(x) = \begin{cases} 1 & \text{if } 0 \leq x \leq 1 \\ 0 & \text{otherwise} \end{cases}, \quad f_{2n}(x) = \begin{cases} 1 & \text{if } 1 < x < 2 \\ 0 & \text{otherwise} \end{cases}$$
$(n = 1, 2, \ldots)$. Show that
$$\int_{-\infty}^{\infty} (\liminf f_n(x))dx = 0 \quad \text{and} \quad \liminf \int_{-\infty}^{\infty} f_n(x)dx = 1.$$
(Thus strict inequality can occur in Fatou's lemma.)

6. If $f \in L^1(\mathbf{R})$, show that
$$\lim_{r \to \infty} \int_{-r}^{r} \left(1 - \frac{|x|}{r}\right) f(x)dx = \int_{-\infty}^{\infty} f(x)dx.$$

7. If φ is a bounded, continuous function on $[0, \infty)$, show that
$$\lim_{t \to 0+} \int_{0}^{\infty} \frac{t\varphi(x)}{t^2 + x^2} dx = \frac{\pi}{2} \varphi(0).$$

8. Show that the function Φ defined by
$$\Phi(x) = \int_{0}^{\infty} \exp\left(-t^2 - \frac{x^2}{t^2}\right) dt \quad (x \in \mathbf{R})$$
is differentiable at each positive real number and
$$\Phi'(x) = -2\Phi(x) \quad (x > 0).$$
Deduce that $\Phi(x) = \frac{\sqrt{\pi}}{2} \exp(-2|x|) \quad (x \in \mathbf{R})$.

9. If $a, b \in \mathbf{R}$, write $a \sim b$ if and only if $(a-b)$ is rational. Then $\sim$ is an equivalence relation on $\mathbf{R}$. If $a \in \mathbf{R}$, denote the $\sim$-equivalence

class of a by $[a]$. Let E be a set obtained by selecting one element from each of the sets $[a] \cap [-\frac{1}{2}, \frac{1}{2}]$. (The existence of such a set depends on a set-theoretic assumption, known as the Axiom of Choice.) Show that E is not measurable.

[Hint: if $r_1, r_2, \ldots$ is an enumeration of the rational numbers in $[-1, 1]$, then the sets $E_n = \{x+r_n : x \in E\}$, $n = 1, 2, \ldots$, are pairwise disjoint and

$$[-\tfrac{1}{2}, \tfrac{1}{2}] \subseteq \bigcup_{n=1}^{\infty} E_n \subseteq [-\tfrac{3}{2}, \tfrac{3}{2}].$$

10. If f is a non-negative function in $L^1(\mathbf{R})$, show that the function λ defined on the set $\mathscr{M}$ of measurable subsets of $\mathbf{R}$ by

$$\lambda(E) = I(f\chi_E) \qquad (E \in \mathscr{M})$$

is a measure.

11. Let v be a measure on a σ-ring $\mathscr{S}$ of sets. Show that
(a) if $A, B \in \mathscr{S}$ and $A \subseteq B$, then $v(A) \leqslant v(B)$ and, provided $v(A)$ is finite, $v(B) - v(A) = v(B-A)$;
(b) if (A_n) is a sequence of sets in $\mathscr{S}$ and $A_n \subseteq A_{n+1}$ ($n = 1, 2, \ldots$), then $v(\bigcup_{n=1}^{\infty} A_n) = \lim v(A_n)$;
(c) if (A_n) is a sequence of sets in $\mathscr{S}$, $A_n \supseteq A_{n+1}$ ($n = 1, 2, \ldots$) and there is a positive integer m such that $v(A_m) < \infty$, then

$$v(\bigcap_{n=1}^{\infty} A_n) = \lim v(A_n).$$

[Hint: in (b), express $\bigcup_{n=1}^{\infty} A_n$ as a countable union of pairwise disjoint sets in $\mathscr{S}$.]

12. Let v be a measure on a σ-ring $\mathscr{S}$ of subsets of a set X. Let L be the collection of functions f on X such that there are real numbers $c_1, \ldots, c_n$ and sets $E_1, \ldots, E_n$ in $\mathscr{S}$ of finite measure with

$$f = \sum_{j=1}^{n} c_j \chi_{E_j}.$$

Show that
(a) L is a vector lattice of functions;
(b) if f is the function in L described in the above equation, then

$I(f)$ is well-defined by

$$I(f) = \sum_{j=1}^{n} c_j v(E_j);$$

(c) I is a Daniell integral on L.

13. Let $f \in L^p(\mathbf{R})$ and, for each $a \in \mathbf{R}$, define f_a as in Problem 3. Show that the function $a \mapsto f_a$ of $\mathbf{R}$ into $L^p(\mathbf{R})$ is uniformly continuous.

14. Let $f \in L^1(\mathbf{R})$ and g be a bounded, measurable function on $\mathbf{R}$. Show that
(a) for each $x \in \mathbf{R}$, the function $y \mapsto f(x-y)g(y)$ belongs to $L^1(\mathbf{R})$;
(b) the function $f * g$ defined by

$$(f*g)(x) = \int_{-\infty}^{\infty} f(x-y)g(y)\,dy \quad (x \in \mathbf{R})$$

is bounded and uniformly continuous on $\mathbf{R}$.

15. Show that the conclusions of Problem 14 are also valid if $f \in L^p(\mathbf{R})$ and $g \in L^q(\mathbf{R})$ where $p > 1$ and $\frac{1}{p} + \frac{1}{q} = 1$.

16. Let $f \in L^p(\mathbf{R})$ and $g \in L^q(\mathbf{R})$ where $p > 1$ and $\frac{1}{p} + \frac{1}{q} = 1$. Show that
(a) if f, g have compact supports, $f * g$ has compact support;
(b) $\lim_{x \to \infty} (f*g)(x) = 0 = \lim_{x \to -\infty} (f*g)(x)$.

17. Show that the set of functions of the form $\varphi_1 + i\varphi_2$, where φ_1, φ_2 are step functions on $\mathbf{R}$, is dense in $L^1(\mathbf{R})$.

18. Let $f \in L^1(\mathbf{R})$. Show that, for each $t \in \mathbf{R}$, the function g defined by

$$g(x) = f(x)e^{-itx} \quad (x \in \mathbf{R})$$

belongs to $L^1(\mathbf{R})$. Show further that the function $\hat{f}$ defined by

$$\hat{f}(t) = \int_{-\infty}^{\infty} f(x)e^{-itx}\,dx \quad (t \in \mathbf{R})$$

is uniformly continuous on $\mathbf{R}$ and $\lim_{t \to \infty} \hat{f}(t) = 0 = \lim_{t \to -\infty} \hat{f}(t)$.

19. Show that there is a function F in $L^1(\mathbf{R}^2)$ which is positive at each point of $\mathbf{R}^2$.

20. Show that every continuous function on $\mathbf{R}^2$ is measurable.

21. If E is a null subset of $\mathbf{R}$, show that $E \times \mathbf{R}$ is a null subset of $\mathbf{R}^2$.

22. Let $F \in L^1(\mathbf{R}^2)$. According to Fubini's theorem, $F(\cdot, y) \in L^1(\mathbf{R})$ for a.a. y in $\mathbf{R}$. Show that there may exist numbers y for which $F(\cdot, y)$ is not even measurable.

23. If E is a null subset of $\mathbf{R}$, show that $E_1 = \{(x, y) \in \mathbf{R}^2 : x - y \in E\}$ is a null subset of $\mathbf{R}^2$.
[Hint: consider the sets $E_1 \cap S_m$, $m = 1, 2, \ldots$, where S_m is the square $\{(x, y) \in \mathbf{R}^2 : -m \leqslant x, y \leqslant m\}$.]

24. If $F(x, y) = \exp(-xy^2) \sin x$, show that for any $t > 0$
$$\int_0^t \left(\int_0^\infty F(x, y) \, dy \right) dx = \int_0^\infty \left(\int_0^t F(x, y) \, dx \right) dy,$$
and deduce that
$$\int_0^\infty \frac{\sin x}{\sqrt{x}} \, dx = \lim_{t \to \infty} \int_0^t \frac{\sin x}{\sqrt{x}} \, dx = \sqrt{\frac{\pi}{2}}.$$

25. Show that the repeated integrals $\int_0^1 \left(\int_0^1 \frac{x^2 - y^2}{(x^2 + y^2)^2} \, dx \right) dy,$

$\int_0^1 \left(\int_0^1 \frac{x^2 - y^2}{(x^2 + y^2)^2} \, dy \right) dx$ both exist, but are not equal. Why does this not contradict Tonelli's theorem?

CHAPTER 5
FOURIER TRANSFORMS

§28. L^1 theory: elementary results

In this chapter, we shall discuss the theory of Fourier transforms, one of the most important tools in applied mathematics. This discussion will provide incidentally some rather nice applications of our work in Chapter 4. We begin by defining the Fourier transform of a function in $L^1 = L^1(\mathbf{R})$ and establishing some elementary properties of such transforms.

28.1 DEFINITION. The Fourier transform $\hat{f}$ of a function f in L^1 is defined by

$$\hat{f}(t) = \int_{-\infty}^{\infty} f(x) e^{-itx} dx \quad (t \in \mathbf{R}).$$

($\hat{f}$ is sometimes defined by

$$\hat{f}(t) = \frac{1}{\sqrt{(2\pi)}} \int_{-\infty}^{\infty} f(x) e^{-itx} dx \quad (t \in \mathbf{R}).$$

As we shall see later, this definition is more convenient in the L^2 theory of Fourier transforms. However, the appearance of factors $\dfrac{1}{\sqrt{(2\pi)}}$ throughout the L^1 theory is a considerable nuisance.)

28.2 THEOREM. Let $f \in L^1$. Then $\hat{f}$ is uniformly continuous on $\mathbf{R}$ and
$$\lim_{t \to \infty} \hat{f}(t) = 0 = \lim_{t \to -\infty} \hat{f}(t).$$
(See Problem 4.18.)

28.3 Theorem. *Let $f, g \in L^1$ and $\alpha \in \mathbf{C}$. Then*
(a) $(\alpha f)\hat{\ } = \alpha \hat{f}$ *and* $(f+g)\hat{\ } = \hat{f} + \hat{g}$;
(b) $(f * g)\hat{\ } = \hat{f}\hat{g}$;
(c) $\sup_{t \in \mathbf{R}} |\hat{f}(t)| \leq \|f\|_1$.

Proof. (a) is obvious. To prove (b) and (c), we note that, for each $t \in \mathbf{R}$,

$$(f*g)\hat{\ }(t) = \int_{-\infty}^{\infty} (f*g)(x) e^{-itx} dx$$

$$= \int_{-\infty}^{\infty} \left(\int_{-\infty}^{\infty} f(x-y) g(y) dy \right) e^{-itx} dx$$

$$= \int_{-\infty}^{\infty} g(y) \left(\int_{-\infty}^{\infty} f(x-y) e^{-itx} dx \right) dy$$

by Tonelli's theorem

$$= \int_{-\infty}^{\infty} g(y) \left(\int_{-\infty}^{\infty} f(u) e^{-it(u+y)} du \right) dy$$

$$= \left(\int_{-\infty}^{\infty} f(u) e^{-itu} du \right) \left(\int_{-\infty}^{\infty} g(y) e^{-ity} dy \right)$$

$$= \hat{f}(t) \hat{g}(t)$$

and

$$|\hat{f}(t)| = \left| \int_{-\infty}^{\infty} f(x) e^{-itx} dx \right| \leq \int_{-\infty}^{\infty} |f(x)| dx = \|f\|_1.$$

(b) and (c) follow.

28.4 Theorem. *Let f be a differentiable function on $\mathbf{R}$ such that f' is continuous and $f, f' \in L^1$. Then*

$$(f')\hat{\ }(t) = it\hat{f}(t) \text{ for each } t \in \mathbf{R}.$$

(The hypothesis that f' is continuous is unnecessary for the conclusion. It has only been made to simplify the proof.)

Proof. We first note that $\lim_{x \to \infty} f(x) = 0$. For suppose this is not the case. Then there is an $\varepsilon > 0$ such that for every real number x_0, there is a real number x_1 with $x_1 \geq x_0$ and $|f(x_1)| \geq \varepsilon$. Since $f' \in L^1$, there is an x_0 in $\mathbf{R}$ such that $\int_{x_0}^{\infty} |f'(x)| dx < \frac{1}{2}\varepsilon$. Choose $x_1 \geq x_0$ with $|f(x_1)| \geq \varepsilon$. Then, for all $v \geq x$.

so that
$$|f(y)-f(x_1)| = \left|\int_{x_1}^{y} f'(x)dx\right| \leq \int_{x_0}^{\infty} |f'(x)|\,dx < \tfrac{1}{2}\varepsilon,$$
$$|f(y)| \geq |f(x_1)| - |f(y)-f(x_1)| > \tfrac{1}{2}\varepsilon.$$

This contradicts the hypothesis that $f \in L^1$. Thus $\lim_{x \to \infty} f(x) = 0$. Similarly $\lim_{x \to -\infty} f(x) = 0$. It follows that, for each $t \in \mathbf{R}$,

$$\begin{aligned}(f')\hat{\,}(t) &= \int_{-\infty}^{\infty} f'(x)e^{-itx}dx \\ &= [f(x)e^{-itx}]_{-\infty}^{\infty} + it\int_{-\infty}^{\infty} f(x)e^{-itx}dx \quad \text{(integrating by parts)} \\ &= it\int_{-\infty}^{\infty} f(x)e^{-itx}dx \\ &= it\hat{f}(t).\end{aligned}$$

§29. The inversion theorem

To illustrate one of the ways in which the above results can be applied, we look at a very simple example. Let g be a continuous function in L^1, and consider the problem of finding the function f such that $f, f', f'' \in L^1, f''$ is continuous and

$$f'' - f = g. \qquad (29.1)$$

If f satisfies (29.1), then, taking Fourier transforms, we see that, for each $t \in \mathbf{R}$, $(f'')\hat{\,}(t) - \hat{f}(t) = (it)^2\hat{f}(t) - \hat{f}(t) = \hat{g}(t)$ i.e.

$$\hat{f}(t) = -\frac{\hat{g}(t)}{1+t^2}. \qquad (29.2)$$

To complete the solution of the problem, we need to know that (29.2) determines f uniquely and, more particularly, to have a formula expressing f in terms of $\hat{f}$. It is the purpose of this section to establish such an inversion formula.

It is convenient to consider the function H defined by

$$H(x) = e^{-|x|} \quad (x \in \mathbf{R})$$

and the functions h_λ, $\lambda > 0$, defined by

$$h_\lambda(x) = \frac{1}{2\pi} \int_{-\infty}^{\infty} H(\lambda t) e^{ixt} dt \quad (x \in \mathbf{R})$$

$$= \frac{1}{2\pi} \left(\int_{-\infty}^{0} e^{(\lambda+ix)t} dt + \int_{0}^{\infty} e^{(-\lambda+ix)t} dt \right)$$

$$= \frac{1}{2\pi} \left(\frac{1}{\lambda+ix} - \frac{1}{-\lambda+ix} \right)$$

$$= \frac{\lambda}{\pi(\lambda^2+x^2)}.$$

We note that, for each $\lambda > 0$, $h_\lambda \in L^1$ and

$$\int_{-\infty}^{\infty} h_\lambda(x) dx = \frac{1}{\pi} \left[\arctan \frac{x}{\lambda} \right]_{-\infty}^{\infty} = 1. \tag{29.3}$$

We require several preliminary results about these functions.

29.1 LEMMA. *Let $f \in L^1$ and $\lambda > 0$. Then*

$$(f * h_\lambda)(x) = \frac{1}{2\pi} \int_{-\infty}^{\infty} H(\lambda t) \hat{f}(t) e^{ixt} dt \quad (x \in \mathbf{R}).$$

Proof. We note that the function h_λ is bounded and measurable, and so $f * h_\lambda$ is defined everywhere on **R**. (See Problem 4.14.) For each $x \in \mathbf{R}$, we have

$$(f * h_\lambda)(x) = \int_{-\infty}^{\infty} f(x-y) h_\lambda(y) dy$$

$$= \frac{1}{2\pi} \int_{-\infty}^{\infty} f(x-y) \left(\int_{-\infty}^{\infty} H(\lambda t) e^{iyt} dt \right) dy$$

$$= \frac{1}{2\pi} \int_{-\infty}^{\infty} H(\lambda t) \left(\int_{-\infty}^{\infty} f(x-y) e^{iyt} dy \right) dt$$

$$ \text{by Tonelli's theorem}$$

$$= \frac{1}{2\pi} \int_{-\infty}^{\infty} H(\lambda t) \left(\int_{-\infty}^{\infty} f(u) e^{i(x-u)t} du \right) dt$$

$$= \frac{1}{2\pi} \int_{-\infty}^{\infty} H(\lambda t) \hat{f}(t) e^{ixt} dt.$$

29.2 LEMMA. *Let g be a bounded, measurable function on $\mathbf{R}$. Then* $\lim_{\lambda \to 0+} (g * h_\lambda)(x) = g(x)$ *at each point x at which g is continuous.*

Proof. Suppose g is continuous at x. Then, for each $\lambda > 0$,

$$(g * h_\lambda)(x) - g(x) = \int_{-\infty}^{\infty} [g(x-y) - g(x)] h_\lambda(y) \, dy \quad \text{by (29.3)}$$

$$= \int_{-\infty}^{\infty} [g(x-y) - g(x)] \frac{h_1(y/\lambda)}{\lambda} \, dy$$

$$= \int_{-\infty}^{\infty} [g(x - \lambda u) - g(x)] h_1(u) \, du$$

by a simple change of variable. Since

$$|g(x - \lambda u) - g(x)| \, |h_1(u)| \leq 2 \sup_{y \in \mathbf{R}} |g(y)| \, |h_1(u)|, \quad h_1 \in L^1$$

and

$$\lim_{\lambda \to 0+} [g(x - \lambda u) - g(x)] h_1(u) = 0 \quad \text{for each } u \in \mathbf{R},$$

the result follows by the dominated convergence theorem.

29.3 LEMMA. *Let $f \in L^p$ where $p \geq 1$. Then*

$$\lim_{\lambda \to 0+} \|f * h_\lambda - f\|_p = 0.$$

(At the moment, we only require this result for $p = 1$; in the next section, however, it will be useful to have the result for $p = 2$.)

Proof. If $\lambda > 0$, $p > 1$ and $\frac{1}{p} + \frac{1}{q} = 1$, then h_λ, being a bounded function in L^1, belongs to L^q and $f * h_\lambda$ is defined everywhere on $\mathbf{R}$. (See Problem 4.15.) Further, for each $x \in \mathbf{R}$,

$$|(f * h_\lambda)(x) - f(x)|^p = \left| \int_{-\infty}^{\infty} [f(x-y) - f(x)] h_\lambda(y) \, dy \right|^p \quad \text{by (29.3)}$$

$$\leq \left(\int_{-\infty}^{\infty} |f(x-y) - f(x)| \, h_\lambda(y)^{1/p} h_\lambda(y)^{1/q} dy \right)^p$$

$$\leq \int_{-\infty}^{\infty} |f(x-y) - f(x)|^p h_\lambda(y) \, dy$$

by Hölder's inequality

so that

$$\|f*h_\lambda - f\|_p^p \le \int_{-\infty}^{\infty} \left(\int_{-\infty}^{\infty} |f(x-y)-f(x)|^p h_\lambda(y)\,dy \right) dx$$

$$= \int_{-\infty}^{\infty} \left(\int_{-\infty}^{\infty} |f(x-y)-f(x)|^p dx \right) h_\lambda(y)\,dy$$

$$= \int_{-\infty}^{\infty} \|f_{-y}-f\|_p^p h_\lambda(y)\,dy \qquad (29.4)$$

where f_{-y} is defined as in Problem 4.3. Clearly this inequality is also satisfied when $p = 1$.

Define the function g on **R** by

$$g(y) = \|f_y - f\|_p^p \quad (y \in \mathbf{R}).$$

According to Problem 4.13, g is continuous on **R**. Also

$$0 \le g(y) \le \{\|f_y\|_p + \|f\|_p\}^p = 2^p \|f\|_p^p \text{ for each } y \in \mathbf{R},$$

so that g is bounded. Hence, by 29.2,

$$\lim_{\lambda \to 0+} \int_{-\infty}^{\infty} \|f_{-y}-f\|_p^p h_\lambda(y)\,dy = \lim_{\lambda \to 0+} (g*h_\lambda)(0) = g(0) = 0.$$

(29.4) therefore implies the result.

29.4 Theorem. *Let $f, \hat{f} \in L^1$. Then, for a.a. x in **R**,*

$$f(x) = \frac{1}{2\pi} \int_{-\infty}^{\infty} \hat{f}(t) e^{ixt}\,dt; \qquad (29.5)$$

in particular, (29.5) is satisfied at any point x at which f is continuous.

(We note that this inversion theorem remains valid if the hypothesis $\hat{f} \in L^1$ is dropped, provided the integral $\frac{1}{2\pi} \int_{-\infty}^{\infty} \hat{f}(t) e^{ixt}\,dt$ is interpreted as the limit $\lim_{s \to \infty} \frac{1}{2\pi} \int_{-s}^{s} \hat{f}(t) e^{ixt}\,dt$. We shall not attempt to prove this generalization.)

Proof. By 29.1,

$$(f*h_\lambda)(x) = \frac{1}{2\pi} \int_{-\infty}^{\infty} H(\lambda t)\hat{f}(t) e^{ixt}\,dt \quad (x \in \mathbf{R}).$$

Since $\lim_{\lambda \to 0+} H(\lambda t)\hat{f}(t)e^{ixt} = \hat{f}(t)e^{ixt}$, $|H(\lambda t)\hat{f}(t)e^{ixt}| \le |\hat{f}(t)|$ for each $\lambda > 0$ and, by hypothesis, $\hat{f} \in L^1$, the dominated convergence theorem implies that

$$\lim_{\lambda \to 0+} (f * h_\lambda)(x) = \frac{1}{2\pi} \int_{-\infty}^{\infty} \hat{f}(t) e^{ixt} dt \quad (x \in \mathbf{R}).$$

According to 29.3, $\lim_{\lambda \to 0+} \|f * h_\lambda - f\|_1 = 0$. Hence, by 26.7, there is a sequence (λ_n) of positive real numbers, converging to 0, such that

$$\lim (f * h_{\lambda_n})(x) = f(x) \quad \text{for a.a. } x \text{ in } \mathbf{R}.$$

Thus (29.5) is satisfied for a.a. x in $\mathbf{R}$.

Finally, if f is continuous at x, then (29.5) is satisfied; for otherwise, since the function g defined by

$$g(y) = \frac{1}{2\pi} \int_{-\infty}^{\infty} \hat{f}(t) e^{iyt} dt \quad (y \in \mathbf{R})$$

is continuous,

$$\{y \in \mathbf{R} : |f(y) - g(y)| > \tfrac{1}{2} |f(x) - g(x)|\}$$

is a neighbourhood of x and is therefore not a null set, a contradiction.

29.5 COROLLARY. *If $f, g \in L^1$ and $\hat{f}(t) = \hat{g}(t)$ for each $t \in \mathbf{R}$, then $f = g$ a.e.*

Proof. If $\hat{f} = \hat{g}$, then $(f-g)\hat{} = 0$ and so, by 29.4, we see that $f(x) - g(x) = 0$ for a.a. x in $\mathbf{R}$.

§30. L^2 theory

$L^2 = L^2(\mathbf{R})$ is not a subset of L^1, and so 28.1 does not define the Fourier transform of every function in L^2. However, if $f \in L^1 \cap L^2$, then $\hat{f}$ is given by 28.1. We shall see presently that, in this case, $\hat{f} \in L^2$ and $\|\hat{f}\|_2 = \sqrt{(2\pi)} \|f\|_2$. It follows that the function $T: f \mapsto \hat{f}$ of $L^1 \cap L^2$, considered as a subspace of L^2, into L^2 is continuous. Since $L^1 \cap L^2$ is a dense subset of L^2 and the space L^2 is complete, T can be uniquely extended to a continuous function, which we again denote by T, of L^2 into L^2: in fact, the extension T maps L^2 onto itself. If $f \in L^2$, Tf is called the **Fourier** (or **Plancherel**) **transform** of f. The Fourier transform, defined on L^2, is of fundamental importance in the mathematical foundations of quantum mechanics.

Above we have given a brief resumé of the L^2 theory of Fourier transforms. We now fill in the details.

30.1 Theorem. *Let* $f \in L^1 \cap L^2$. *Then* $\hat{f} \in L^2$ *and*

$$\|\hat{f}\|_2 = \sqrt{(2\pi)} \, \|f\|_2. \qquad (30.1)$$

(If $\hat{f}$ had been defined by $\hat{f}(t) = \dfrac{1}{\sqrt{(2\pi)}} \int\limits_{-\infty}^{\infty} f(x)e^{-itx} dx$ $(t \in \mathbf{R})$, then (30.1) would become $\|\hat{f}\|_2 = \|f\|_2$.)

Proof. Define $\tilde{f}$ on $\mathbf{R}$ by

$$\tilde{f}(x) = \overline{f(-x)} \quad (x \in \mathbf{R}).$$

Clearly $\tilde{f} \in L^1 \cap L^2$. Hence $g = \tilde{f} * f$ is a bounded, continuous function in L^1. By 29.1,

$$(g * h_\lambda)(0) = \frac{1}{2\pi} \int\limits_{-\infty}^{\infty} H(\lambda t)\hat{g}(t) dt$$

for each $\lambda > 0$, where H, h_λ are the functions introduced in §29. By 29.2,

$$\lim_{\lambda \to 0+} (g * h_\lambda)(0) = g(0) = \int\limits_{-\infty}^{\infty} \tilde{f}(-y)f(y) dy = \|f\|_2^2.$$

For each $t \in \mathbf{R}$,

$$\hat{g}(t) = (\tilde{f})\hat{\ }(t)\hat{f}(t) = \left(\int\limits_{-\infty}^{\infty} \tilde{f}(x)e^{-itx} dx\right)\hat{f}(t) = |\hat{f}(t)|^2 \geq 0$$

and $H(\lambda t)$ increases to 1 as λ decreases to 0, i.e.

$$H(\lambda_1 t) \geq H(\lambda_2 t) \quad \text{if} \quad 0 < \lambda_1 < \lambda_2 \quad \text{and} \quad \lim_{\lambda \to 0+} H(\lambda t) = 1.$$

Hence, by the monotone convergence theorem, $\hat{g} = |\hat{f}|^2 \in L^1$, i.e. $\hat{f} \in L^2$, and

$$\lim_{\lambda \to 0+} \frac{1}{2\pi} \int\limits_{-\infty}^{\infty} H(\lambda t)\hat{g}(t) dt = \frac{1}{2\pi} \int\limits_{-\infty}^{\infty} \hat{g}(t) dt = \frac{1}{2\pi} \|\hat{f}\|_2^2.$$

(30.1) follows.

30.2 Theorem. *The function* $T: f \mapsto \hat{f}$ *of* $L^1 \cap L^2$, *considered as a subspace of* L^2, *into* L^2 *is continuous and can be uniquely extended to a continuous function, again denoted by* T, *of* L^2 *into itself.*

Proof. If $f, g \in L^1 \cap L^2$, then, by (30.1),

$$\| Tf - Tg \|_2 = \| (f-g)\hat{\ } \|_2 = \sqrt{(2\pi)} \| f-g \|_2.$$

It follows that $T: L^1 \cap L^2 (\subseteq L^2) \mapsto L^2$ is continuous.

Let $f \in L^2$. Since $L^1 \cap L^2$ is dense in L^2 (see 26.8), there is a sequence (f_n) in $L^1 \cap L^2$ such that

$$\lim \| f - f_n \|_2 = 0. \tag{30.2}$$

Since $\| Tf_p - Tf_q \|_2 = \sqrt{(2\pi)} \| f_p - f_q \|_2$, the sequence (Tf_n) in L^2 is Cauchy, and, as L^2 is complete, therefore converges to g, say. g is independent of the sequence (f_n) in $L^1 \cap L^2$ satisfying (30.2): for if (g_n) is a second sequence in $L^1 \cap L^2$ satisfying (30.2), then

$$\| Tg_n - g \|_2 \leq \| Tg_n - Tf_n \|_2 + \| Tf_n - g \|_2$$
$$= \sqrt{(2\pi)} \| g_n - f_n \|_2 + \| Tf_n - g \|_2 \to 0.$$

Thus we can define $Tf = g$, thereby extending T to L^2.

The extension T of L^2 into itself is continuous. For if $f \in L^2$ and (f_n) is a sequence in $L^1 \cap L^2$ converging to f, then

$$\| Tf \|_2 = \lim \| Tf_n \|_2 = \sqrt{(2\pi)} \lim \| f_n \|_2 = \sqrt{(2\pi)} \| f \|_2, \tag{30.3}$$

so that T is continuous at 0. As T is easily shown to be linear, it follows that T is continuous at each point of L^2.

Finally, if U is a continuous extension of $T: L^1 \cap L^2 (\subseteq L^2) \mapsto L^2$ and $f \in L^2$, say f is the L^2-limit of the sequence (f_n) in $L^1 \cap L^2$, then $\lim \| Tf_n - Uf \|_2 = \lim \| Uf_n - Uf \|_2 = 0$ and $Uf = Tf$.

The equation $\| Tf - Tg \|_2 = \sqrt{(2\pi)} \| f-g \|_2$ $(f, g \in L^1 \cap L^2)$ shows that the function T of $L^1 \cap L^2$, considered as a subspace of L^2, into L^2 is uniformly continuous. Thus we could have deduced 30.2 immediately from Problem 2.13.

30.3 PARSEVAL'S IDENTITY. *Let $f, g \in L^2$. Then*

$$\int_{-\infty}^{\infty} f(x) \overline{g(x)} \, dx = \frac{1}{2\pi} \int_{-\infty}^{\infty} (Tf)(x) \overline{(Tg)(x)} \, dx.$$

Proof. $f\bar{g} = \frac{1}{4}\{|f+g|^2 - |f-g|^2 + i|f+ig|^2 - i|f-ig|^2\}$, and there is a similar formula for $Tf\overline{Tg}$. Hence, by (30.3),

$$\int_{-\infty}^{\infty} f(x)\overline{g(x)}\,dx$$
$$= \tfrac{1}{4}\{\|f+g\|_2^2 - \|f-g\|_2^2 + i\|f+ig\|_2^2 - i\|f-ig\|_2^2\}$$
$$= \frac{1}{8\pi}\{\|Tf+Tg\|_2^2 - \|Tf-Tg\|_2^2 + i\|Tf+iTg\|_2^2 - i\|Tf-iTg\|_2^2\}$$
$$= \frac{1}{2\pi}\int_{-\infty}^{\infty} (Tf)(x)\overline{(Tg)(x)}\,dx.$$

30.4 Theorem. (a) *The function T of L^2 into L^2, defined in 30.2, is one-one and maps L^2 onto itself;*
(b) *the relation between $f \in L^2$ and $g = Tf$ can be described in the following way: if, for each t in $\mathbf{R}$,*

$$g_n(t) = \int_{-n}^{n} f(x)e^{-itx}\,dx, \quad f_n(t) = \frac{1}{2\pi}\int_{-n}^{n} g(x)e^{itx}\,dx, \quad (30.4)$$

then

$$\lim \|g - g_n\|_2 = 0 \quad \text{and} \quad \lim \|f - f_n\|_2 = 0.$$

Proof. (a) (30.3) implies that T is one-one. To show that T has range L^2, consider the function U defined on $L^1 \cap L^2$ by

$$(Uf)(t) = \frac{1}{2\pi}\int_{-\infty}^{\infty} f(x)e^{itx}\,dx \quad (f \in L^1 \cap L^2,\ t \in \mathbf{R}).$$

If $f \in L^1 \cap L^2$, then $(Uf)(t) = \dfrac{1}{2\pi}(Tf)(-t)$ for each t in $\mathbf{R}$, so that $Uf \in L^2$ and $\|Uf\|_2 = \dfrac{1}{\sqrt{(2\pi)}}\|f\|_2$. As in 30.2, U can be uniquely extended to a continuous function, again denoted by U, of L^2 into itself. If $f, Tf \in L^1 \cap L^2$, then, by the inversion theorem, 29.4, $UTf = f$ or, interchanging the roles of T, U,

$$TUf = f. \quad (30.5)$$

Let $f \in L^1 \cap L^2$. Then, for each $\lambda > 0$, $f * h_\lambda \in L^1 \cap L^2$ and, according to 30.1, $T(f * h_\lambda) = \hat{f}\hat{h}_\lambda \in L^1 \cap L^2$. Further, by 29.3, $\lim_{\lambda \to 0+} \|f * h_\lambda - f\|_2 = 0$. Hence, since T, U are continuous functions of L^2 into itself, we have in L^2

$$TUf = \lim_{\lambda \to 0+} TU(f * h_\lambda) = \lim_{\lambda \to 0+} f * h_\lambda \quad \text{by (30.5)}$$
$$= f.$$

Since $L^1 \cap L^2$ is dense in L^2, (30.5) must, in fact, be satisfied for all f in L^2. It follows that T has range L^2.

(b) Let $f \in L^2$, $g = Tf$ and f_n, g_n be defined by (30.4). Then $g_n = T(f\chi_{[-n,n]})$ and, since the sequence $(f\chi_{[-n,n]})$ converges in L^2 to f, we see that $Tf = g$ is the L^2-limit of (g_n). Also $f_n = U(g\chi_{[-n,n]})$ and, since the sequence $(g\chi_{[-n,n]})$ converges in L^2 to g, we see that $Ug = f$ is the L^2-limit of (f_n).

Problems on Chapter 5

1. Let $f \in L^1$ and suppose the function g defined by
$$g(x) = xf(x) \quad (x \in \mathbf{R})$$
also belongs to L^1. Show that $\hat{f}$ is differentiable at each point t of $\mathbf{R}$ and $(\hat{f})'(t) = -i\hat{g}(t)$.

2. If $\delta > 0$, show, using the inversion theorem, that the function h defined by
$$h(x) = \begin{cases} 1 - \dfrac{|x|}{2\delta} & \text{for } |x| \leqslant 2\delta \\ 0 & \text{otherwise} \end{cases}$$
is the Fourier transform of the function g in L^1 defined by
$$g(x) = \frac{1}{\delta\pi} \frac{\sin^2 \delta x}{x^2}.$$
Deduce that if F is a closed subset of $\mathbf{R}$ and $x_0 \in \mathbf{R} - F$, there is a function g_0 in L^1 such that
$$\hat{g}_0(x_0) \neq 0 \text{ and } \hat{g}_0(x) = 0 \text{ for each } x \in F.$$

3. Show that there is no function u in L^1 such that
$$f * u = f \text{ for all } f \in L^1.$$

4. Let (u_n) be a sequence of non-negative functions in L^1 such that
(a) u_n has compact support $[a_n, b_n]$ where $\lim a_n = 0 = \lim b_n$;
(b) $\int_{-\infty}^{\infty} u_n(x) dx = 1.$

Show that $\lim \| f * u_n - f \|_1 = 0$ for each $f \in L^1$. (A sequence (u_n) of functions in L^1 satisfying this condition is called an **approximate identity** for L^1.)

5. For each positive integer n, write

$$u_n(x) = \frac{1}{n\pi} \frac{\sin^2 nx}{x^2} = v_n(x) + w_n(x),$$

$$v_n(x) = \begin{cases} u_n(x)(1 - |x|\sqrt{n}) & \text{for } |x| \leq \frac{1}{\sqrt{n}} \\ 0 & \text{otherwise} \end{cases}$$

Show that

(a) $\int_{-\infty}^{\infty} v_n(x) dx = 1 - r_n$ where $r_n > 0$ and $\lim r_n = 0$;
(b) (v_n) is an approximate identity for L^1;
(c) $\lim \| f * w_n \|_1 = 0$ for each $f \in L^1$.
Deduce that (u_n) is an approximate identity for L^1.

[Hint: in (a), use the fact that $\dfrac{1}{\pi} \int_{-\infty}^{\infty} \dfrac{\sin^2 x}{x^2} dx = 1$. See Problem 2.]

6. Use Problems 2 and 5 to show that the set S of functions f in L^1 such that $\hat{f}$ has compact support is dense in L^1.